Geographic Knowledge Infrastructure

Geographic Knowledge Infrastructure

Applications to Territorial Intelligence and Smart Cities

Robert Laurini

ELSEVIER

First published 2017 in Great Britain and the United States by ISTE Press Ltd and Elsevier Ltd

ISTE Press Ltd
27-37 St George's Road
London SW19 4EU
UK

www.iste.co.uk

Elsevier Ltd
The Boulevard, Langford Lane
Kidlington, Oxford, OX5 1GB
UK

www.elsevier.com

Notices

For information on all our publications visit our website at http://store.elsevier.com/

British Library Cataloguing-in-Publication Data
A CIP record for this book is available from the British Library
Library of Congress Cataloging in Publication Data
A catalog record for this book is available from the Library of Congress
ISBN 978-1-78548-243-4

Printed and bound in the UK and US

Contents

Preface

In 1989, I co-authored a small book in French regarding spatial knowledge engineering [LAU 89] with my colleague and friend, Françoise Milleret-Raffort. It was the beginning of a long research journey (Figure 1) on applying information technologies to environmental and urban planning. Then in 1993, with Derek Thompson, I wrote *Fundamentals of Spatial Information Systems* [LAU 93], which was well received by the GIS community, and rapidly became a best-seller in GIS, not only for students in geography and in computer science but also GIS researchers and specialists. The third book [LAU 01], *Information Systems for Urban Planning*, was more devoted to urban problems and applications.

Figure 1. *The three books*

Now, after the so-called Artificial Intelligence winter, research in knowledge engineering is constantly growing. Territorial Intelligence and Smart Cities are now very common buzzwords used to highlight the future: this future is not shaped by urban planners and politicians as much as it is by the people or citizens interested in it. Presently, as the philosopher Michel Serres said, knowledge must be considered as an infrastructure.

The big problem is then how to amalgamate artificial intelligence and human collective intelligence to shape a better life for our children? This book tries to give a few answers by installing preliminary pillars to bridge this gap between those very different types of intelligence. But there is still a long way to go!

In this book, overall 2D knowledge will be studied. Of course, urban planning and territorial intelligence also need 3D and 3D+Time reasoning. Integrating 2D reasoning in those domains is considered the first step to deal with upper dimensions.

In the first chapters of this book, fundamental elements for territorial intelligence, smart city and knowledge representation methods are given in order to sketch the structure and the contents of geographic knowledge bases. Then, geographic objects, relations, structures, ontologies and gazetteers are studied as the core components.

After a few elements regarding geographic knowledge discovery through data mining, geographic rules are presented as the pivot for studying consequences about any environmental or urban projects. Novel geovisualization techniques are then studied in order to give decision-makers in local authorities new tools to understand the aftermath of their policies visually.

The last chapter presents concepts relative to knowledge infrastructure for territorial intelligence and smart city governance in order to gain some further insight.

I would like to thank all people who have helped me carry out this research and especially all those who invited me all over the world to give seminars, conferences and to organize panels by allowing me to present and validate my ideas. I also want to acknowledge all those who have not hesitated to ask difficult questions, thus generating novel developments in geographic knowledge engineering.

Οἱ κύκνοι, ἐπειδὰν αἴσθωνται ὅτι δεῖ αὐτοὺς ἀποθανεῖν, ᾄδοντες καὶ ἐν τῷ πρόσθεν χρόνῳ, τότε δὴ πλεῖστα καὶ κάλλιστα ᾄδουσι, γεγηθότες ὅτι μέλλουσι παρὰ τὸν θεὸν ἀπιέναι οὗπέρ εἰσι θεράποντες (From Plato, *Phaedo* 85a).

Robert LAURINI
January 2017

1

From Geodata
to Geographic Knowledge

In many domains, it is more and more common to speak about knowledge, wherefrom the expression "knowledge society" was coined. According to *Wikipedia*[1], "A knowledge society generates, shares and makes available to all members of the society, knowledge that may be used to improve the human condition". Starting from this definition, it could be interesting to examine how knowledge can improve not only the management of a city or a territory but also urban planning at large. The goal of this introductory chapter is to give a few directions for renovating urban and environment planning through knowledge engineering.

Of course, humans are the focus of territorial intelligence, but the use of knowledge technologies can help amplify human reasoning not only by generating new development scenarios and studying alternatives of urban development, but also evaluating the consequences in various terms, human, societal, financial, etc.

After some historical reminders, several definitions concerning smart cities and territorial intelligence will be analyzed in order to sketch smart urban and territorial governance. In the third part, we will examine the promises of knowledge engineering technologies for renovating urban and regional planning.

1 https://en.wikipedia.org/wiki/Knowledge_society.

1.1. A rapid history of urban planning and information technology

For practically half a century, information technology has profoundly transformed urban planning. Initially, it was only true for some statistics and mathematical modeling of cities [BAX 75, BAX 76] and then for map-making. During these years, the expression "computer-assisted cartography" was used, emphasizing how computers could help the automatic creation of maps. At the end of the 1970s, it became obvious that automatic cartography would have to be seen differently, and geographic or urban data was then stored into databases. The expression "GIS" (Geographic Information Systems) was coined for software systems able not only to store geographic data and make maps but also those integrating tools devoted to spatial analysis. Applications are so numerous that it is impossible to draw up a complete list; the expression "from Archeology to Zoology" is a way to show the coverage of potential applications [LAU 01]. Associations such as the "Urban Data Management Society[2]" have organized symposia since 1970 in Europe in order to promote information technology not only for urban planning, but also for city government. See also URISA[3] in the US.

As a consequence, urban planning has gradually been renovated, first by data (modeling and statistics), then by information (especially GIS and spatial analysis) and now by knowledge. It will be the role of this book to explain how to use knowledge and construct knowledge base systems to renovate urban planning and city management.

1.2. Territorial intelligence, smart cities and smart planning

Many definitions have been proposed to define both smart cities and territorial intelligence. They all have in common the integration of sustainable development. Even though those two concepts come from different disciplines, their main differences can be stated as follows: "Smart City" focuses on cities, "Territorial Intelligence" concerns larger spaces including regions and countries.

---

2 http://www.udms.net.
3 http://www.urisa.org/.

1.2.1. *Smart cities*

Dr. Carlo Ratti, director of the MIT Senseable City Lab, claims that an intelligent or smart city is technological, interconnected, clean, attractive, comforting, efficient, open, collaborative, creative, digital and green. The European Union considers six components (Figure 1.1): economy, mobility, environmental, people, living and governance to shape a smart city.

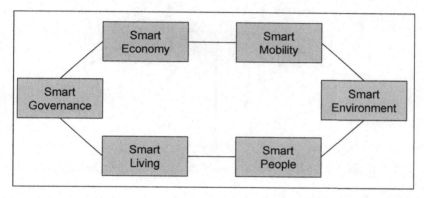

Figure 1.1. *Definition of smart cities according to the European Union*

Boyd Cohen[4] has given a more sophisticated definition of a smart city with this circular diagram or wheel (Figure 1.2), inspired by the work of many others. In this diagram, one can see that Information and Communication Technologies (ICT) appear several times especially under the title "Smart Government".

This latter definition was extended by Mathew [MAT 13], illustrated in Figure 1.3 as a form of a diamond connecting Smart Governance, Smart Citizens, Smart Healthcare, Smart Energy, Smart Buildings, Smart Technology, Smart Infrastructures and Smart Mobility. Remark "Smart Infrastructures" in this diamond!

---

4 http://www.fastcoexist.com/1680538/what-exactly-is-a-smart-city.

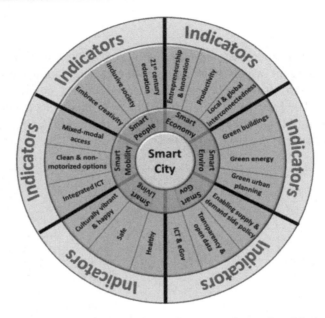

Figure 1.2. *Definition of a Smart City according to Boyd Cohen*

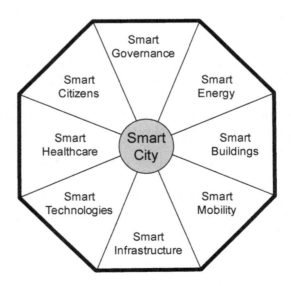

Figure 1.3. *The Smart City components according to Mathew's diamond [MAT 13]*

In addition, let me mention another interesting definition emphasizing participation: according to Deutsche Telecom[5], "a Smart City is an ecosystem characterized by a partially digitized set of processes and striving to its self-awareness and efficiency, through ICT and a higher degree of participation from its citizens, authorities and businesses". Moreover, according to [KOU 12], "Smart cities are the result of knowledge-intensive and creative strategies aiming at enhancing the socio-economic, ecological, logistic and competitive performance of cities. Such smart cities are based on a promising mix of human capital (e.g. skilled labor force), infrastructural capital (e.g. high-tech communication facilities), and social capital (e.g. intense and open network linkages) and entrepreneurial capital (e.g. creative and risk-taking business activities".

Note that the last definition stresses the importance of knowledge in a smart city. For other definitions and analysis, please refer to [ALB 15] for a very comprehensive review.

Recently [AIR 16] proposed the definition of an "intelligent urban ecosystem" combining smart people, smart governance and smart businesses (Figure 1.4).

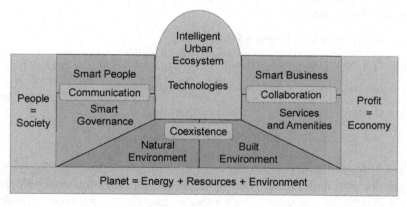

Figure 1.4. *Another view of intelligent urban ecosystem. Modified from [AIR 16]*

5 Cited by http://www.cisco.com/assets/global/RO/events/2015/ciscoconnect/pdf/santal/Cisco-Introductory-EP.pdf.

Another interesting point of view is given by [FER 16] in which she compares 32 definitions from a stakeholder point of view. In conclusion, she proposes the following definition: "A Smart City is a system that enhances human and social capital wisely using and interacting with natural and economic resources via technology-based solutions and innovation to address public issues and efficiently achieve sustainable development and a high quality of life on the basis of a multi-stakeholder, municipally based partnership". This definition can be seen much closer to territorial intelligence.

1.2.2. *Territorial intelligence*

Considering territorial intelligence, also several definitions can be quoted. According to [BER 07], "Territorial Intelligence can be compared with the territoriality which results from the phenomenon of appropriation of resources of a territory; it consists of know-how transmissions between categories of local actors of different cultures". On the other hand, [GIR 08] defines Territorial Intelligence as "the science having for object the sustainable development of territories and having for subjects the territorial communities". This definition was extended later [GIR 10] by specifying that territorial intelligence innovations must include:

– use of multidisciplinary knowledge;

– dynamic vision of territories;

– involvement of communities and practitioners;

– sharing, co-constructing and cooperating;

– and participatory territorial governance.

Territorial intelligence consists of a set of scientific methodologies, analysis tools and measurement systems that mobilize the stakeholders of a determined territory. The territorial intelligence approach entails coordinating all stakeholders to bring about actions favoring the collective good (citizens, companies, etc.) through an interaction of the three concepts of sustainable development: economy, society and environment (Figure 1.5).

For my part, let me propose the following definition: "Territorial Intelligence can be defined as an approach regulating a territory (maybe a city) which is planned and managed by the cross-fertilization of human collective intelligence and artificial intelligence for its sustainable development".

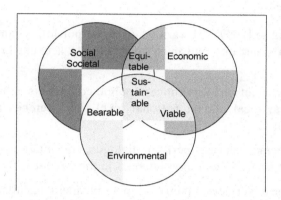

Figure 1.5. *The three pillars for territorial intelligence*

And now the question [LAU 14, LAU 15b] is "how artificial intelligence and especially knowledge engineering can help not only local decision-makers to plan a city but also lay citizens to give their opinion about the future of their city?". A preliminary study can be found in [BER 16].

1.2.3. *Smart Urban Planning*

Smart Urban Planning can be seen as a possible answer. Similarly, several visions are possible, among others [ANT 12] and [FIS 13], by having all three facets, sustainable development, greater involvement of citizens and major use of technologies.

By examining the difference between the words smart and intelligent, the authors of [FIS 13] explain that the adjective intelligent seems to imply the capability of developing actions in order to solve a problem by using methods and information contained in a knowledge base whereas the word smart seems to have, apart from the cognitive heritage (even if not organized in an analytical way), also the power of solving the problem "operatively", showing which are the "tools" to be used for the specific purpose. Summing up, while the intelligent thinks, works out and suggests the models to adopt in order to find a solution, the smart shows also the operative way and the devices to use.

According to [ANT 12], various *e*-service portfolios can be offered in a modern smart city and must be taken into account in smart city planning:

– *e*-government services concern public complaints, administrative procedures at local and at national level, job searches and public procurement;

– *e*-democracy services perform dialogue, consultation, polling and voting about issues of common interests in the city area;

– *e*-business services mainly support business installation, while they enable digital market places and tourist guides;

– *e*-health and tele-care services offer distant support to particular groups of citizens such as the elderly, civilians with diseases, etc.;

– *e*-learning services offer distant learning opportunities and training material to the habitants;

– *e*-security services support public safety via amber-alert notifications, school monitoring, natural hazard management, etc.;

– environmental services contain public information about recycling, while they support households and enterprises in waste/energy/water management. Moreover, they deliver data to the State for monitoring and for decision making on environmental conditions such as for microclimate, pollution, noise, traffic etc. (in Ubiquitous and Eco-city approaches);

– Intelligent Transportation supports the improvement of the quality of life in the city, while it offers tools for traffic monitoring, measurement and optimization;

– communication services such as broadband connectivity, digital TV, etc.

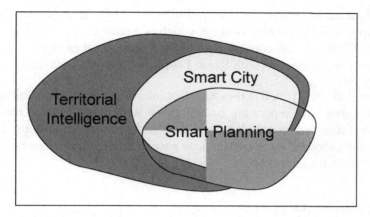

Figure 1.6. *Main differences between Smart Cities, Territorial Intelligence and Smart Planning*

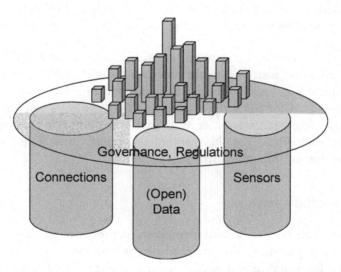

Figure 1.7. *The three pillars of the Smart City from [MUR 15]*

But what are the main differences between Smart City, Territorial Intelligence and Smart Planning (Figure 1.6)? The first difference is that Smart City only deals with cities whereas Territorial Intelligence deals with larger territories. The second difference is that Territorial

Intelligence integrates the human dimension while Smart City can be accompanied by another concept which is Smart People. Concerning Smart Planning, this concept regards also some other aspects of urban planning in connection with Smart People and Smart Governance.

For their part, [MUR 15] explain that information and communication technologies are supporting smart cities with three pillars, namely sensors, data and especially open data and connections (Figure 1.7). Regarding sensors, some explanations are needed before they can become mainstream for smart cities.

1.3. Data acquisition sensors

As exemplified in Figure 1.7, one of the pillars of a smart city is a set of sensors. By sensor, one means a device able to measure a phenomenon and to transform it into digital values. Initially designed for physical phenomena, their definition has been recently extended.

In other words, the sensors must measure a phenomenon, be connected to an instrument that picks up the signal and translates it into a reading in engineering units, connected to a recorder that will store the signal for further processing. Usually, sensors can send data regularly, whether directly or not, to some monitoring center, for instance each hour, but this center can modify this frequency. In addition, sensors are often GPS-positioned.

1.3.1. *Digital sensors*

Digital sensors can measure temperature, pressure, humidity, noise, air or water pollution, the presence of gas, motion and acceleration, etc. (see for instance *Wikipedia*[6] for a complete list of sensors); the majority of them can be purchased by less than US $10.

---

6 https://en.wikipedia.org/wiki/Sensor.

From an architectural point of view, various types of sensors exist (Figure 1.8):

– those which can be connected directly to a computer with a cable (Figure 1.8(a)), for instance a USB cable; some can be connected to a gateway;

– those which can send data and those which can also receive data through wireless communications (Figure 1.8(b)) in connection with a monitoring center, with some gateway or even with other sensors;

– and those with some storage capacity (Figure 1.8(c)).

When there are many sensors sending signals to others, a sort of telecommunication route can be defined as exemplified in Figure 1.8(d).

Now, imagine a city with thousands or millions of sensors transmitting their data wirelessly and regularly: the level of the electrosmog will be strong. According to some specialists, they recommend using cables, especially optic fiber, in the future to reduce electrosmog.

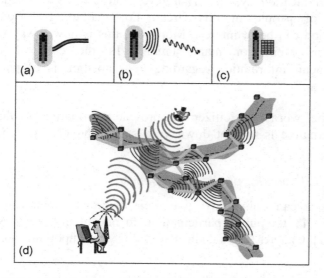

Figure 1.8. *Sensors: a) sensor with wired communications;
b) wireless sensor; c) sensor with storage capacity;
d) concept of telecommunication route*

An extension of the sensor system is known as the Internet of Things which is a novel architecture in which everything is connected to everything. This consideration is outside the scope of this book, although the Internet of Things[7] can be a major component of a smart city.

1.3.2. *Citizens as sensors – crowdsourcing*

In his seminal paper, Goodchild [GOO 07] claimed that citizens can be considered as sensors voluntarily or involuntarily. Voluntarily, when they can send information to some monitoring office, or involuntarily because by having their smart phone in their pocket, they can be traced.

1.3.2.1. *Crowdsourcing*

By definition, crowdsourcing is a mode of data acquisition in which anybody can supply. When you are walking in a city with your smart phone in your pocket, your position is known by the telecommunication system. Therefore, you can be counted in the number of people in this street. For marketing reasons, this information can be of interest. Moreover, this is now a way to count people in events or manifestations. In the case of disaster management, information regarding your position is of paramount importance.

In other words, any citizen can provide information involuntarily. The alternative is to shut down your smart phone or to leave it at home.

1.3.2.2. *Citizens for public participation*

Citizens can voluntarily give their opinions which can be the basis of GIS for public participation. In one of my previous books [LAU 01], Chapter 9 was dedicated to this very important topic. Even

---

7 https://en.wikipedia.org/wiki/Internet_of_Things.

if a lot of works and experiences have been carried out in many places, public participation is still a challenge.

But within the scope of this book, the question must be revisited: to what degree do citizens exhibit geographic knowledge or located good practices which can be machine-processable? As stakeholders, they have some logics. It is very clear that NIMBY (Not-In-My-Backyard) people are fighting for their own limited interest. But the real question is: are they individuals who have a clear vision of global interest? Can they help in constructing an urban plan? Can their logic be used for reasoning about potential consequences?

1.4. About reasoning

In the computing literature, it is common to see the expression "spatial reasoning". But what is spatial reasoning and how is it different from geographic reasoning? In this section, some answers will be provided.

1.4.1. *Spatial reasoning*

In domains such as robotics, spatial reasoning is common; for instance when moving a robot in a room given a list of obstacles. Sometimes, spatial reasoning is another word for geometric reasoning and the solutions are generally obtained by computational geometry. There are many problems with nice names, such as the piano-movers, problem that is how to move a geometric object with strange shape in a messy place. There is also the knapsack problem emphasizing how to put the maximum number of objects into a bag, etc. More common, the Sudoku game is a kind of spatial reasoning.

The solution is generally obtained through logic, but in connection with geometry and topology.

1.4.2. *Geographic reasoning*

According to Carsten Braun[8], here are the core components of geographic reasoning:

– "measuring and mapping geographic distributions (where is stuff?);

– identifying patterns and clusters (stuff usually does not occur randomly on Earth!);

– identifying paths and flows (think about roads and traffic!);

– analyze these geographic relationships (over time, if applicable)".

This list is restricted to geographers in general. For us, we need to include the analysis of consequences of any possible decisions of local authorities in urban and environmental management and planning.

1.5. Promises of geographic knowledge

As previously enumerated, geographic knowledge must be multidisciplinary. One of the ways to represent knowledge is by using rules. In planning, the rules have the following origins: physical (water, floods, vegetation, landslides, etc.), societal (economy, etc.), administrative (laws, decrees, etc.) or even from best practices. In addition, other rules can be extracted from spatial data mining [LAU 15b, LAU 16b].

One of the difficulties is the fact that among the urban actors, some have different "logics". With regard to industrial zoning, an environmentalist or an industrialist may have different ideas on the possible implications of this or that choice. Similarly, some groups may have different priorities: facing an empty space, athletes imagine a stadium, pupil's parents a school and a realtor a building, etc. From

8 http://carstenbraun.blogspot.fr/2013/05/what-is-geographic-reasoning.html.

a formal point of view, these aspects will occur in multi-actor and multi-criteria decision support systems.

In their paper dated more than 25 years ago, Batty and Yeh [BAT 91] exposed the promises of the so-called expert systems in urban planning. However, in that period, those systems were only built on logic, very difficult to use, with limited interfaces. Now, 25 years after, with the evolution of computer science, information technologies, artificial intelligence and geovisualization, new approaches can be integrated to design new kinds of intelligent systems especially devoted to geographic applications and overall urban planning. Let us call them geographic inference engines which will be able to make reasoning about geographic knowledge. Whereas a conventional inference engine is only based on logics, such a novel system must integrate topology, computational geometry, statistics and spatial analysis because geographic rules necessitate those aspects to be modeled.

According to Reginald Golledge [GOL 02], Geographic Knowledge is useful for two fundamental reasons: (1) to establish where things are and (2) to remember where things are to help us in the process of making decisions and solving social and environmental problems. However, some questions and tasks that the discipline must face as we move further into the 21st Century include:

– how can geographic knowledge contribute to the comprehension and solution of problems involved in society-space relations?

– what future role can geographic knowledge play in establishing global international, national, regional, and local policy?

– what geographic knowledge can we create to enhance understanding of global societies, cultures, economies, and political and information structures?

It will be the scope of this book to try to provide not only partial answers, but also directions in order to reach Gollege's objectives.

In his controversial essay, Lacoste [LAC 76] explains that the initial goal of geography is war. Of course, this definition could be applied to Alexander the Great, Hannibal, Julius Caesar, Napoléon, Hitler, etc. But other definitions can be found, such as for Henry the Navigator, the scope of geography was not only for exploring and discovering the world but also conquering new territories with commerce as an ultimate goal. Consider also famous explorers such as Christopher Columbus, Vasco de Gama, Ferdinand Magellan, Roald Amundsen, etc.

Bearing all that in mind, concerning territorial intelligence, let me propose the following definition: "geographic knowledge corresponds to information potentially useful to explain, manage, monitor and plan a territory". Let me develop:

a) *Geographic knowledge to explain.* This corresponds more or less to Gollege's definition. Synonyms can be to understand, to explore, to assess the context and to detect problems. Existing books and monographs can help a lot from a historical point of view. Techniques such as geographic text mining [SAL 13] and, when databases are existing, spatial data mining (see Chapter 9), can be the sources of this kind of geographic knowledge.

b) *Geographic knowledge to manage.* One of the goals of local authorities is to manage the territory under their jurisdiction. The management could range from street and engineering network repairs to school and other public services such as waste collection. The knowledge they have to use is essentially coming from laws, by-laws and best practices. In other words, knowledge is known in some natural language and must be transformed to become machine-processable. Here, can knowledge be seen as an extension to business intelligence applied to local authorities?

c) *Geographic knowledge to monitor.* This kind of knowledge can be seen as an extension of the previous one, but its nature is totally different. Indeed, in order to reduce pollution or regulate traffic, local authorities install sensors as previously explained to obtain raw data which are transformed into knowledge by real-time data mining.

d) *Geographic knowledge to plan.* To my understanding, this is the ultimate goal of geographic knowledge engineering, to plan smart cities or territories, meaning to design scenarios of evolution, study alternatives and take citizen's opinions into account within the scope of sustainable development.

As a consequence, two types of knowledge can be distinguished:

– *low-level knowledge* concerns the transformation of raw data into geographic objects and the relations between them. It includes also (*i*) quality control, (*ii*) extracted knowledge coming from database data mining, (*iii*) gazetteers, and (*iv*) independence from data storage location. It integrates also issues such as: independence from data acquisition techniques, ontologies; independence from scales and resolution; independent from geographic representations and basic algorithms, and possibly integrating multilingualism aspects.

– *high-level knowledge* regards physical laws (hydrology, demographics, pollution, etc.), juridical laws, best practices, linked to applications and also documents.

1.6. Conclusion: advocacy for geographic knowledge infrastructures

In [EDW 13], knowledge infrastructures are defined as "robust networks of people, artifacts, and institutions that generate, share, and maintain specific knowledge about the human and natural worlds".

Consequently, based on this background, it is possible to define a geographic knowledge infrastructure. For decades, governments, whether national or local, have developed spatial data infrastructures. Similarly, it is possible to envision geographic knowledge infrastructure (Figure 1.9) as bunches of knowledge necessary to developed higher level applications, those bunches coming either from data mining over the spatial data infrastructure[9] or from human collective intelligence able to be formalized in a machine-processable format. Of course, many bunches cannot yet be formalized.

---

9 Many papers deal with the design and use of Spatial Data Infrastructures. A good synthesis is given in https://en.wikipedia.org/wiki/Spatial_data_infrastructure.

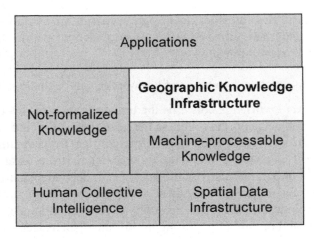

Figure 1.9. *Geographic knowledge infrastructure*

Chunks of low-level knowledge will be directly detailed in subsequent chapters dealing with geographic objects and relations, whereas high-level knowledge will be studied in the geographic rule chapter (Chapter 10) as a basis for territorial intelligence, and smart city planning and governance.

2

Knowledge Representation

How can we represent knowledge? What are the main differences between data, information and knowledge? Why is knowledge important in human reasoning? What are the specificities of spatial and geographic reasoning? What are the different forms of knowledge? How can we encode chunks of knowledge? What could be the promises of automatic geographic reasoning? Those are a few questions which will be tackled in this chapter.

Knowledge representation of the world (in Artificial Intelligence) is the ability to build a model of objects, their links and the actions they can perform. Knowledge representation is the expression in a well-form language, a model that expresses the knowledge about the world. There are three common ways: (*i*) declarative description of the state of the world, (*ii*) procedural expression of the transformations of states and (*iii*) an object-oriented description of existing objects. As the second is used overall in programming, in this book, only the first and the third will be developed.

For many mathematicians and computer scientists, knowledge representation and reasoning is based on logic, often under the expression of "deductive reasoning". However, for roboticists, spatial reasoning is based on computational geometry and topology.

In this chapter, the objective will be to examine all these aspects. After having given a few hints about the present possibilities in automatic reasoning, various tools coming from logic will be detailed.

2.1. Automated reasoning

Reasoning is the ability to make inferences, and automated reasoning is concerned with the building of computing systems that automate this process. Although the overall goal is to mechanize different forms of reasoning, the term has largely been identified with valid deductive reasoning as practiced in mathematics and formal logic. In other words, automated reasoning allows humans to amplify their own reasoning by the help of computers. For mathematicians, the big problem is how to ensure the validity of deductions overall by using theorems, without questioning the validity of premises, that is of basic knowledge chunks, objects, relations and rules. But for users, the big problem is to ensure that basic knowledge is valid, without errors or leading to dead ends.

2.1.1. *Data, information and knowledge*

What are the differences between data, information and knowledge? Let us examine two small illustrative examples.

The first deals with traffic lights. "Red" and "Green" are data. "Traffic light is red" is a piece of information. "When the traffic light is red, the driver has to stop" is a chunk of knowledge rule. And more "When I see a red light, I stop", this is wisdom by applying existing rules.

The second example deals with SOS. "Dots" and "dashes" are data. The information relative to SOS is encoded "Three dots, three dashes, three dots". If I hear those raw signals, I know this is an emergency alert. In this case, by deciding to help, this is wisdom.

In other words (Figure 2.1), data are raw signals (bits, numbers, string of characters, etc.). Information corresponds to the meaning attached to data. Knowledge attaches purpose and competence to information and can generate potential action. Another way of defining knowledge is to claim that knowledge is information useful to solve a problem. Using knowledge adequately is the beginning of wisdom. Figure 2.2 shows another extension due to [DAV 98] by claiming that data and information regard business operation whereas knowledge and wisdom concern business intelligence.

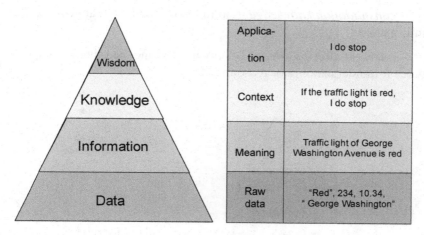

Figure 2.1. *Data, information, knowledge and wisdom*

There have been several papers which suggest how to identify knowledge in geographic and spatial systems. As a synthesis of some previous works, [CRO 99] proposed the following classification:

– *primitive knowledge* about the identification of primitives. A primitive is a readily identifiable point, line or area which cannot be subdivided into smaller named entities; this includes knowledge about an object's size and shape if relevant;

– *relationship knowledge* of the spatial relationships between primitives in terms of their proximity, orientation and degree of overlap;

– *assembly knowledge*, used to define collections of objects which form identifiable spatial decompositions; it includes knowledge of the spatial density of primitives; this knowledge can be regarded as knowledge needed for generalization;

– *non-visual knowledge* which helps refine classifications including labeling of scene primitives and spatial relationships (spatial knowledge); it consists of temporal knowledge, algorithmic knowledge and heuristic knowledge;

– *consolidation knowledge* used to resolve and evaluate conflicting information;

– *interpretation knowledge* of how to combine the other knowledge types for understanding or reasoning.

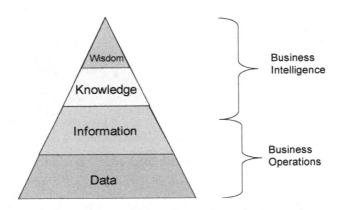

Figure 2.2. *Relations with business operations and intelligence. Adapted from [DAV 98]*

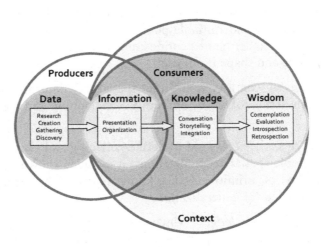

Figure 2.3. *Relationships between Data, Information, Knowledge and Wisdom, according to [SHE 94]*

More explicitly, Nathan Shedroff[1] built the previous diagram (Figure 2.3) illustrating the relationships between data, information, knowledge and wisdom.

2.1.2. *Vision of Turban/Aaronson*

The various origins of knowledge are depicted in Figure 2.4 [TUR 98]. Let us emphasize that not only facts and rules must be stored but also processes, constraints, heuristics, decision rules, procedures for problem solving, etc. In some knowledge bases also typical situations, best practices are stored in order to create an organizational memory.

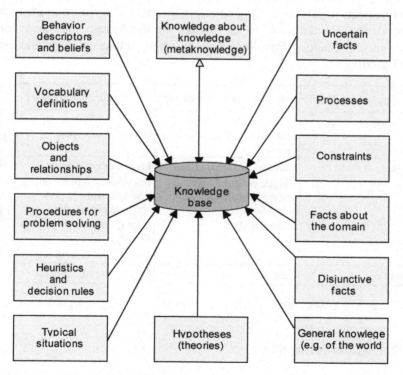

Figure 2.4. *Type of knowledge to be represented in the knowledge base. Made according to [TUR 98] with modifications*

1 http://www.xenogenic.com/viz/i-u-1.0.htm.

A very important distinction can be made between explicit and tacit types of knowledge (see Table 2.1). Tacit knowledge (knowing-how) is knowledge embedded in the human mind through experience and jobs, know-how and learning, personal wisdom and experience, context-specific, more difficult to extract and codify. In addition, tacit knowledge includes insights and intuitions. However, explicit knowledge (knowing-that) is knowledge codified and digitized in books, documents, reports, memos, etc. and also documented information that can facilitate action. Explicit knowledge is what is easily identified, articulated, shared and employed, in a word, machine-processable.

See also the SECI model created by [NON 95]. Regarding geographic knowledge and database, the situation is not very clear. A few aspects can be really explicit whereas the great majority of knowledge chunks look to be hidden in databases. By means of various tools, among them data mining, some chunks can be extracted and used.

Type of knowledge	Implicit (Tacit) knowledge	Explicit knowledge
Objective	Know-how by learning, Embedded within people minds.	Documented information that can facilitate action.
Where to find?	Informal and uncodified. Values, perspectives & culture Knowledge in people's heads. Hidden in databases.	Formal or codified Documents: reports, policy manuals, white papers, standard procedures Books, magazines, journals (library) Issued from data mining
Entities	Sentences	Objects, Relations, Rules

Table 2.1. *Main differences between explicit and implicit knowledge*

2.2. Logical formalisms

The properties of a good representation are representational adequacy, inferential adequacy, inferential efficiency and efficiency in the acquisition/modification. Several different formalisms can follow those recommendations; among them, the more common way is based on logics. Four kinds of logics can be of interest, propositional logic, predicate logic, descriptive logic and fuzzy logic. Let us rapidly give their main characteristics. Indeed, the logical formalisms for knowledge are represented by a set of well-formed formulas (the so-called WFF). The inference mechanisms are based on deductive logic methods. See for instance [MEN 97] for details.

2.2.1. *Propositional logic*

Propositional logic (sometimes called propositional calculus) is based on a language the syntax of which is defined by well-formed formulas and semantics by assigning values to variables. Its formal system is based on language axioms, and its inference rules are complete and consistent. In this context, a proposition is an affirmative assertion which we can say that is true or false (but not both!). Examples of true propositions can be: "Yesterday it rained in Puebla", "The Earth revolves around the sun", "$2 \times 3 = 3 + 3$", "7 is an odd number", etc. Considering false propositions, let us mention "$2 + 3 = 6$", "the sun revolves around the Earth", as a preliminary example, let us consider the following propositions:

– premise 1: "If it is raining then it is cloudy";

– premise 2: "It is raining";

– conclusion: "It is cloudy".

Formally, this can be written

– premise 1: $P \rightarrow Q$;

– premise 2: P;

– conclusion: Q.

Syntactically, let us define a set *PROP* of propositions characterized by the following assertions:

a) $p_i \in PROP$ for any $i \in N$

b) If $\alpha \in PROP$ and $\beta \in PROP$ then

$(\alpha \wedge \beta) \in PROP$ (union)

$(\alpha \vee \beta) \in PROP$ (intersection)

$(\alpha \rightarrow \beta) \in PROP$ (implication)

$(\alpha \leftrightarrow \beta) \in PROP$ (equivalence)

c) If $\alpha \in PROP$ then $(\neg \alpha) \in PROP$ (negation).

There are two different ways to justify the veracity of the hypothesis. Either it implies the truth of the conclusion, that is semantic justification noted $PROP \models P$, or you can reach the conclusion from the hypothesis across duly justified steps, that is syntactic justification, noted $PROP \vdash P$.

Three types of deductions are common:

– *modus ponens*: $(A \rightarrow B, A) \rightarrow B$;

– *modus tollens*: $(A \rightarrow B, \neg B) \rightarrow \neg A$;

– disjunctive syllogism: $(A \vee B, \neg A) \rightarrow B$.

From a set of propositions, it is generally possible to deduce some other propositions provided that the set of propositions is consistent and exhaustive. However, the major big problem is to transform any situation to a set of valid propositions.

2.2.2. Predicate logic

This formal system is characterized by its formulae that contain variables which can be quantified. Two common quantifiers are the existential $\exists$ ("there exists") and universal $\forall$ ("for all") quantifiers. The variables could be elements in the universe under discussion, or perhaps relations or functions over that universe. For instance, consider the sentence: "All dogs are mammals and Rex is a dog, then Rex is a mammal". It can be written:

$\forall x\ Dog(x) \rightarrow Mammal\ (x)$

$Dog\ (Rex) \rightarrow Mammal\ (Rex)$

The veracity of reasoning depends on the relationships among the set of propositions. Remark that propositional logic is not sufficiently expressive to capture this relationship.

Predicate logic is an interesting formalism for knowledge representation. Among the advantages, mention that it has well defined syntax and semantics and can easily handle quantization aspects. Moreover, its complete inference system (can be extended to resolution method) is established. However, among limitations, remark that there are limits on the expressive power (regarding possibilities and uncertainty) and problems regarding the implementation of non-monotonous reasoning.

2.2.3. Descriptive logics

Descriptive logics (DL) are a family of knowledge representation formalisms. They define the relevant concepts in a domain and relate them to specify properties.

The basic syntactic building blocks are the Atomic concepts (unary predicates), Atomic roles (binary predicates) and individuals (constant). The expressive power of language is restricted to a small set of constructors to build complex concepts and roles. The implicit knowledge about concepts and individuals can be inferred automatically with the help of inference procedures.

In this context, a knowledge base includes two "boxes":

– TBox contains sentences describing hierarchical concepts (i.e. interrelations between concepts).

Example: Every employee is a person.

– ABox contains statements indicating where in the hierarchy individuals belong. Example: "Paquita is the boss", "Nice is a French city", "Paquita lives in Nice".

The basic notations are given in Table 2.2, in which C is a concept (class), P is a role (property) and x_i is an individual/nominal.

Constructor	DL Syntax	Example	FOL Syntax
intersectionOf	$C_1 \sqcap \ldots \sqcap C_n$	Human $\sqcap$ Male	$C_1(x) \wedge \ldots \wedge C_n(x)$
unionOf	$C_1 \sqcup \ldots \sqcup C_n$	Doctor $\sqcup$ Lawyer	$C_1(x) \vee \ldots \vee C_n(x)$
complementOf	$\neg C$	$\neg$Male	$\neg C(x)$
oneOf	$\{x_1\} \sqcup \ldots \sqcup \{x_n\}$	{john} $\sqcup$ {mary}	$x = x_1 \vee \ldots \vee x = x_n$
allValuesFrom	$\forall P.C$	$\forall$hasChild.Doctor	$\forall y.P(x,y) \rightarrow C(y)$
someValuesFrom	$\exists P.C$	$\exists$hasChild.Lawyer	$\exists y.P(x,y) \wedge C(y)$
maxCardinality	$\leqslant nP$	$\leqslant$1hasChild	$\exists^{\leqslant n} y.P(x,y)$
minCardinality	$\geqslant nP$	$\geqslant$2hasChild	$\exists^{\geqslant n} y.P(x,y)$

Table 2.2. *Main notations for descriptive logic (DL means descriptive logic and FOL first-order-logic). Source: Ian Horrocks with permission*

a) The TBox includes among others:

– definitions of the concepts and properties (relationships) of the controlled vocabulary;

– statements of axioms of concept or functions;

– inference of relationships, whether it is transitive, symmetric, functional or inverse to another property;

– proof of equivalence if two classes or properties are equivalent among themselves;

– subsumption, which is to check if a concept is more general than other;

– satisfiability, is the problem of checking whether a concept is defined (it is not an empty concept);

– classification, which places a new concept in the right place in a taxonomic hierarchy of concepts;

– logical implication, which is a generic relation is a logical consequence of the statements in the box;

– infer implied claims of property using the transitive property.

b) The ABox includes:

– claims of membership, such as concepts and roles;

– claims of attributes;

– claims of links that capture the above but also to assert outside sources for these tasks;

– instances of consistency control;

– controls of satisfiability, are the conditions of membership of the instance.

c) Important links between TBox and ABox:

– linkages, that if other propositions are implicit in the indicated conditions;

– example of verification, that checks whether a given individual is an instance of (belongs to) a specified concept;

– consistency of knowledge base, which is to verify if all the concepts accepted at least have one instance;

– relations of identity, which consists of determining the equivalence or relationship instances on different sets of data;

– disambiguation, which is resolving the references to the appropriate instance.

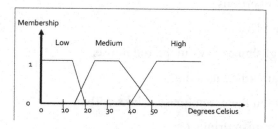

Figure 2.5. *Example of definition of temperatures with fuzzy logic*

2.2.4. *Fuzzy sets and logics*

Who is a senior citizen? Consider a 90-year-old woman; she can be called a senior citizen. But consider a fifty-year old man; is he really a senior citizen? Some will say yes and others no. A solution is found with fuzzy set theory created by [ZAD 65]. In classical set theory, the membership of elements in a set is assessed in binary terms according to a bivalent condition – an element either belongs or does not belong to the set. By contrast, fuzzy set theory permits the gradual assessment of the membership of elements in a set; this is described with the aid of a membership function valued in the real unit interval [0, 1]. With this definition, the old lady will be assigned 100% whereas the man for instance 50%.

Now consider temperatures. What is a low or a high temperature? An answer can be found in Figure 2.5 in which by membership degrees, low, medium and high temperature can be respectively defined.

A lot of mathematical developments have been made regarding fuzzy sets and fuzzy logics. For our concern in geographic knowledge engineering and reasoning, let us mention that fuzzy sets can be of interest. See section 4.4 for details regarding fuzzy geographic objects.

2.2.5. *About Lewis Carroll's example*

The following puzzle corresponds to the description of a context. It was said that Lewis Carroll created the first versions, and now several exist. It can be seen as a very simple spatial problem along a street. Here are the assertions:

1) there are five houses;

2) the Englishman lives in the red house;

3) the Spaniard owns a dog;

4) the man in the green house drinks coffee;

5) the Ukrainian drinks tea;

6) the green house is immediately to the right of the ivory house;

7) the motorcycle owner keeps snails;

8) the man in the yellow house rides a bike;

9) the man in the middle house drinks milk;

10) the Norwegian lives in the first house on the left;

11) the man who rides a skateboard lives next to the man who owns a fox;

12) the man who rides a bike is in the house next to the man who owns a horse;

13) the man who has a hang-glider drinks orange juice;

14) the Japanese man drives a powerboat;

15) the Norwegian lives next to the blue house.

And the questions are: who drinks water? Who owns the zebra?

In order to solve this problem, one can construct the following table (Table 2.3). From assertion #10, we can easily place the Norwegian, then with assertion #9, the milk drinker, etc. As a result, one can deduce that:

– the Japanese owns the zebra;

– and the Norwegian drinks water.

House	1	2	3	4	5
Color	Yellow	Blue	Red	Ivory	Green
Nationality	Norwegian	Ukrainian	Englishman	Spaniard	Japanese
Drink	*Water*	Tea	Milk	Orange juice	Coffee
Vehicle	Bike	Skateboard	Motorcycle	Hang-glider	Powerboat
Pet	Fox	Horse	Snails	Dog	*Zebra*

Table 2.3. *The zebra table to solve a small linear spatial problem*

However, note that in this small example, space is only 1D (namely along a street) and discretized. But in reality, it is 2D or 3D and continuous, the situation is more complex that is to say.

2.2.6. *Logics and space*

The key problem for using logics in geography is the integration of space, namely topology and computational geometry. It is easy to see that we are facing two different visions of mathematics which are of very different origins.

Spatial reasoning can be used by roboticists for example when moving a robot in a 3D room full of objects. But those kinds of reasoning are essentially procedural, that is based on algorithms. Transforming them into a declarative way linked to a spatial inference engine is still a challenge.

In our case, in territorial intelligence (urban and environmental planning), the space can be seen as 2D which looks simpler. All geometric aspects will be treated as special operators or relations which will be included into knowledge chunks. It will be the role of this book to show detail some of them.

2.3. RDF (Resource Description Framework)

RDF[2] as a standard model for data interchange on Internet, is, among other things, a data model built on edge-node "graphs". Each link in a graph consists of three things (with many aliases depending on the mapping from other types of data models) as illustrated in Figure 2.6:

– subject (or start node, instance, entity and feature) is a resource;

---

2 https://www.w3.org/RDF/.

– predicate (or verb, property, attribute, relation, member, link, reference) is a binary relation; types of predicates are identified in the web;

– object (or end node) is a value or a literal.

Any of the three things in a single triple can be represented via a URI (Uniform Resource Identifier), possibly with an optional fragment identifier. Subjects and objects are called nodes and can be represented as a blank node (usually with a local identifier with no meaning). Objects can also be represented as a literal value. Note that the same node may play the role of a Subject in some edges, and the role of the Object in others.

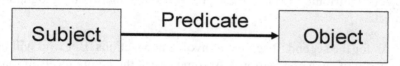

Figure 2.6. *Structure of RDF triples*

Here is presented an example in RDF:

```
<?xml version="1.0"?>
<rdf:RDF
xmlns:rdf=http://www.w3.org/1999/02/22-tdf-
syntax-
  ns# xmlns:cd="http://vww.recshop.fake/cd#">
    <rdf:Description rdf :about=
  "http://www.aa.com/cd/Empire
    Burlesque">
      <cd:artist>Bob Dylan</cd:artist>
      <cd:country>USA</cd:country)
      <cd:company>Columbia</cd:company>
      <cd:price>10.90</cd:price>
      <cd:year>1985</cd:year>
    </rdf:Description>
  </rdf:RDF>
```

2.4. Rule modeling

In artificial intelligence, the representation of rules is based on several mathematical theories, such as classical logics as previously explained. Moreover, according to [GRA 06] and [MOR 08], business rules should be considered first-class citizens in computer science.

2.4.1. *Rules and classical logics*

Generally, their implementation is based on two grammatical structures: IF-THEN-Fact and IF-THEN-Action [ROS 11]. The first serves above all to involve new facts, that is, for us, new objects, attribute values, new relationships between geographic objects.

As for the second, it serves to involve new actions. But who will be in charge of such new actions? In some cases, the computer itself may run procedures; in others, particularly in regulatory contexts, a decision maker (for example, the Mayor of a municipality) must himself initiate the action. Another interpretation could be that the choice of alternatives of an action, for example when a law, in some well-defined contexts, opens many perspectives.

A special case concerns sets of rules related to the same factors-dependent conditions modeled according to trees or decision tables.

2.4.2. *Decision trees and tables*

In preparing decision making, sometimes it is interesting to present the possible solutions with a tree-like structure; it is the called a decision tree. Figure 2.7 gives an illustration of a decision tree. Sometimes, instead of a tree, the possibilities can be regrouped into a table which is called a decision table (Table 2.4). Both decision trees and tables are very common in decision-support systems and they can be seen as special cases of business rules.

Figure 2.7. *Example of a decision tree*

	Rule 1	Rule 2	Rule 3	Rule 4
Conditions				
Frequent Flyer	Gold	Gold	Silver	Silver
Class	Business	Economy	Business	Economy
Actions				
Free upgrade	First	Business	No	Business
Discounted Upgrade	N/A	First	First	First

Table 2.4. *Example of a decision table*

2.4.3. *Rules and fuzzy logic*

In [ISH 05], the authors have proposed a schema to represent fuzzy rules IF-THEN together with a certainty grade (i.e. rule weight) noted *CF* which is a real number between 0 and 1. In this vision, a crisp rule will have *CF*=1. Fuzzy rules can thus be used when that one is dealing with fuzzy geographic objects, but also when the implication is fuzzy such as OFTEN-THEN type. As a result the implication of the classical logical symbol can be understood as ALWAYS-THEN.

This is particularly the case for the modeling of rules such as follows:

– almost all shops are located on the ground floor;

– most Canadians live along the US border;

– rarely a highway passes through a building.

2.4.4. *From business rules to geographic rules*

According to [JOS 10], many business rule engines that are available in the marketplace use rules of the condition–action or event–condition–action types, and both can be executed by a computer. A condition–action type of rule condition–action has the structure:

```
if (some condition is fulfilled) do (perform
some activity)
```

Whereas an event–condition–action rule has the structure:

```
when (some event has happened) and if (some
condition   is   fulfilled)do   (perform   some
activity).
```

Concerning geographic rule, a first idea is to extend conditions by means of a spatial condition prefixed by a Where clause. Chapter 10 will develop this aspect.

2.4.5. *Generic model of rules*

From a mathematical point of view, in this book, rules will be presented as follows (Figure 2.8):

– antecedents will be represented as a context with quantifiers ("$\forall$" or "$\exists$") followed by the symbol ":" and some Boolean expressions to model conditions;

– the symbol "$\Rightarrow$" when the implication is mandatory; more details will be given later for frequent or fuzzy implications;

– and consequents (as acts or actions); if there are many, they will parenthesized by "{" and "}", and each separated by ";". Indeed, some logicians should prefer the symbol "∨", but its semantics are not very clear in this case; indeed the symbol "∨" suggests that some of the consequents are implied, whereas the symbol ";" will means that all consequents must be implied.

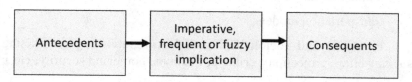

Figure 2.8. *Succession of the components of a rule*

However, from a computing point of view, [BOL 10] suggested several XML extensions to model rules. The simplest of these is as follows:

```
<Implies>
<if>
<..>
</if>
<then>
<..>
</then>
</Implies>
```

Chapter 10 will examine the semantics of geographic applicative rules and propose a more generic model.

2.5. About mathematical models

A lot of knowledge chunks regarding geographic phenomena are modeled in a procedural way, that is with computer programs. Among others, let us mention:

– physical models to describe hydrological aspects such as floods, pollution issues (noise, air pollution, fire propagation, landslides, avalanches, CO^2 emissions, solar irradiation, energy-related models (smart grids), etc.);

– demographic models in order to make projections for the future;

– animal habitat or biotope evolutions;

– transportation models;

– economic and sociological models (migrations, geomarketing, social welfare projections, crime projections, homeland security, etc.);

– etc.

All have in common the use of a lot of data, having been written in procedural languages (C++, Java or even FORTRAN) based on cellular automata, multi-agent systems, differential equations, etc. They can be encapsulated to be transformed so that they could be used in knowledge rules. A special repository in geographic knowledge base systems should be designed to store them. Sometimes they are called what-if models and this name suggest that they can be included in the rules.

2.6. Case-based reasoning

A very common approach to solve a problem is when we know the solution of a similar problem, the clue is to adapt this solution to the problem we have to solve. In other words, we can re-use some preliminary experience. Case-based reasoning is a set of solutions based on this approach. Four steps must be considered:

– retrieve a similar case in a case base;

– reuse this case by adapting it;

– modify this solution if necessary;

– and store the result into the case base.

When all cases are described with the same set of n attributes, the cases can be represented in an n-dimensional space. Hence similar cases can be retrieved by using k^{th}-nearest neighbor algorithms.

In this book, we will not construct any geographic case base, but this approach can be used when an interesting solution is found in one place and one wants to apply it, possibly with modification, in another place (for instance, from External Knowledge issued from technological watching, see section 3.3).

2.7. Conclusion: what is special for geographic knowledge?

The scope of this chapter was to give the basic elements for representation of knowledge in the objective to create efficient systems for territorial intelligence. But there are some difficulties to represent geographic knowledge. Among them, let us mention:

– reality is complex and several mathematical theories must be used to get relevant models; to describe parcels, Euclidean geometry is excellent whereas fuzzy sets can be of interest to describe mountains; and for hydrology, differential equations have proved their efficiency; moreover, for some issues, one has to take Earth rotundity into account;

– for decades or maybe centuries, geometry and logics have been two separated mathematical disciplines and the creation of a joint theory (integrating geometry and logics) is a not so easy challenge;

– topology as a sort of intermediary between logics and geometry could constitute a promising starting point especially for some aspects of spatial reasoning;

– conventional geographic reasoning is often based on spatial analysis and operation research; it will be interesting also to try to integrate them also;

– a lot of knowledge chunks in geography have been encapsulated into numerical models written in a procedural way; a methodology is necessary to transform them into declarative knowledge;

– for reasoning, rules are the key-elements; hence how to pass from geodata with their own characteristics to geographic knowledge chunks and more how to introduce them into geographic rules?

– and finally, each actor has its own vision. By extending what we say in Italian *"ad ognuno la sua verità"*, we can state that each stakeholder has his/her own knowledge, preoccupation and rules. Any geographic reasoning system must integrate this diversity and propose some kind of compromise.

In the following chapters, some answers will be provided to all these questions.

3

Towards Geographic Knowledge Systems

After having sketched the main ideas regarding territorial intelligence, smart cities and knowledge representation, it is time to examine how to construct geographic knowledge bases. In recent decades, Geographic Information Systems (GIS) have shown significant progress, but there are still drawbacks and limitations often resulting from the philosophy behind those systems, especially the underlying assumptions.

The scope of this chapter will be first to examine the lessons learnt by designing and using GIS to give a list of requirements. The principles that will be used in structuring geographic knowledge bases (GKB) will then be presented preceded by a set of preliminary considerations, namely prolegomena.

3.1. Lessons learnt from GIS

Since the end of the 1970s, Geographic Information Systems have been in use. From all these experiences, some lessons have been learnt. The scope of this section will be to analyze the advantages, drawbacks and limitations we are still facing regarding GIS. They include, among others, mathematical visions of the world, the relationships between geographic objects modeling, geovisualization and interoperability.

3.1.1. *About the mathematical visions of the world*

It is common to state that there are 0D, 1D, 2D and 3D objects, but geographic objects in our world are 3D moving objects (3D+T). The usage of the lower dimensions (0D, 1D, 2D) has shown its efficiency in a lot of situations. Apart the North and the South poles, there are no 0D objects in our world even if the summits of the mountains can be often modeled as 0D objects. Regarding real 1D geographic objects, we can only mention the Equator, the Meridians, the Parallels, the Tropic of Cancer, the Tropic of Capricorn, the Arctic Circle and the Antarctic Circle.

The consequence is that the discourse regarding dimensions of geo-objects must be renovated. In this book dealing with territorial intelligence and smart city planning, the mainstream vision is 2D. So the idea is to base our system on 2D objects by giving emphasis to areas. Although it is simple to consider points as small areas, for lines it is a little bit different. In GIS applications, objects such as rivers or roads are often modeled as linear objects often with a width. But theoretically speaking in geometry, lines have no width. In this case, the concept of ribbon will supersede lines, as explained in Chapter 4.

A very important aspect when modeling a geographic feature deals with mathematical representations usually taken as attributes. For years, several models (for instance for storing a simple polygon) have existed, but standardization has opted for only one of them (OGC[1]). For a street, several models exist depending from actor's vision. Figure 3.1 illustrates four families of models totally different:

– the first model is based on graph with edges as street axes and nodes as crossroads;

– the second based on two polylines delimiting the private part from the public part (cadaster meaning);

– the third as an area-based model for describing the section reserved for traffic;

– and finally, a 3D model in order to integrate underground engineering networks.

---

1 http://www.opengeospatial.org/.

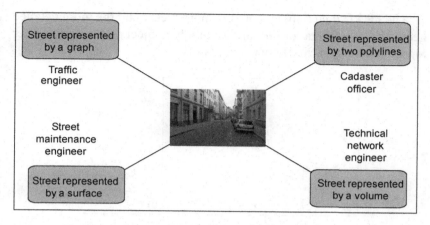

Figure 3.1. *Multiple geometric representations of a street*

The main consequence is that there are difficulties to pass from one model to another, and sometimes some procedures demand a type of model the user does not have. Hence, a unique model must be provided together with procedures for adapting the model when necessary.

3.1.2. *About geo-objects modeling and storing*

Remember that in earlier times (1960s or 1970s), people spoke about Computer-Aided Cartography, so emphasizing cartographic aspects. Then it became apparent that storing geographic information was more important than making or remaking maps possibly with variants. The difficulty in managing different representations induces the idea of multi-representation [HAN 04], that is by storing different representations of the same objects, each of them being used at different scales.

But with algorithms, geometric object generalization can be carried out [DOU 73], possibly on-the-fly. Others [LBA 00] have proposed to distinguish models for storing from models for mapping.

The idea is then only to store the more recent and the more detailed geometric representations of any geographic object and to derive other representations on-the-fly when necessary.

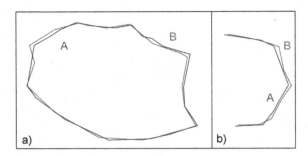

Figure 3.2. *Two homologous geometric descriptions of the same geographic objects: a) polygon homology; b) line homology*

3.1.3. *About data quality and homologies*

Consider two database fragments in which the same city has two geometric descriptions which are generally a little bit different, say for instance two polygons describing the same place but made from different acquisition devices, and we want to compare then. From mathematical and computing points of view, they are not equal, but for humans, they look equal with a certain tolerance.

Between two objects, *A* and *B*, a homology relation is a relation that is reflexive and symmetric which defines a sort of similarity between two things. Let us denote this relation by ₪, so that one can write *A* ₪ *B*. Therefore, both (*A* ₪ *B*) and (*B* ₪ *A*) hold. Remark that an equivalence relation (≡) is a homology relation, which is also transitive. Figure 3.2(a) illustrates two homologous polygons and Figure 3.2(b) two homologous lines. More details will be given in section 4.7.2.

3.1.4. *About multiple representations and granularity of interest*

It is well known that "cartographic representation is linked to visual acuity". Thresholds must be defined. In classical cartography, the limit ranges from 1 mm to 0.1 mm. If one takes a road and a certain scale and if the transformation gives a width of more than 1 mm, this road is an area; then between 1 mm and 0.1 mm, a line; and if less than 0.1 mm the road disappears. The same reasoning is valid for cities or small countries such as Andorra, Liechtenstein, Monaco, etc. In these cases, the "holes" or small islands in Italy or in France disappear cartographically.

A solution based on multiple representations was commonly found in some GIS. In other words, for the same geographic features, various geometric representations are stored. For instance, a country is represented successively by several polygons respectively with 100 points, 1,000 points and 10,000 points. And according to the necessary scale, one representation is selected [ZHO 03, HAN 04].

This solution is generally considered "heavy" to maintain. Another track is to store the polygons with the highest level of detail and to generalize it on-demand. For an excellent survey about geometric generalization, refer to [SPI 95].

The great drawback of the previous definitions is that they are too closely linked to cartography. Instead, let us speak of granularity of interest.

3.1.5. *Requirements for geographic knowledge systems*

In order to create a well-done geographic system, there are some requirements to follow:

– offering a relevant and possibly complete representation of reality;

– offering a robust and accurate representation for any granularity of interest;

– storing consistent and validated knowledge;

– updating regularly;

– supporting geographic reasoning;

– representing any stakeholder's logics;

– combining GKB coming from different sources;

– defining planning projects and assessing them.

3.2. GKS structure

Working with a GKS for local authorities implies to define two levels of knowledge:

– *generic knowledge* which is valid everywhere and linked to acquisition devices and linguistics aspects; the majority of rules given in this book falls into this category;

– *applicative knowledge* which is linked to applicative domains such as urban planning, environmental planning, transportation, etc.; it will be the scope of Chapter 10 to give some hints concerning geographic applicative rules.

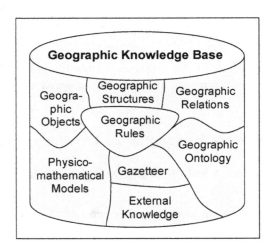

Figure 3.4. *Contents of a geographic knowledge base*

As a consequence, any geographic knowledge base (Figure 3.4) will consist of (details will be given in the subsequent chapters):

– an ontology describing the geographic objects, their relationships between them and their attributes under the form of a graph;

– a gazetteer giving the list of place names (or toponyms), their peculiarities, their variants, etc.;

– a set of geographic objects with their spatial type and attributes in accordance with the ontology and sometimes a toponym as given in the gazetteer (remember that many features do not bear a name, for instance, electricity poles);

– a set of relations between geographic objects which are not mentioned in the ontology;

– some physico-mathematical models especially to describe the geographic phenomena and some other aspects such as land use and transportation; sometimes they are called what-if models;

– a set of geographic rules which will use the vocabulary of the ontology and place names described in the gazetteer and sometimes some mathematical models;

– and a set of bunches representing external knowledge (See section 3.3 for details).

We must add a set of ongoing projects into the GKS but they constitute knowledge with a different status which will be examined later on (section 4.8). Moreover, it can be of interest to capture a set of metadata concerning the whole knowledge base; but this point is outside the scope of this book

In this list, we can add documents which can be considered to be storing geographic information given in natural language. See [SAL 03] for more details.

Figure 3.5 illustrates the links of a geographic knowledge base. Thanks to geographic data mining, some knowledge chunks can be extracted from a database. Similarly, thanks to the techniques of geographic information retrieval, documents can be analyzed.

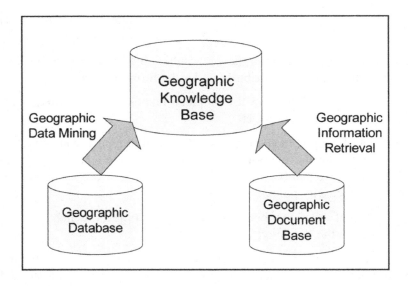

Figure 3.5. *Links between a geographic knowledge base, a geographic database and some geographic documents*

In addition, Figure 3.6 illustrates the links between a geographic object and the gazetteer and the ontology.

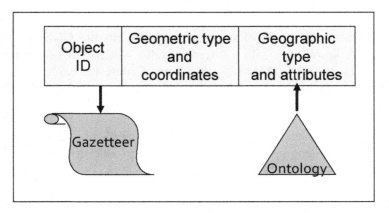

Figure 3.6. *A geographic object with its links with gazetteer and ontology*

Moreover, a geographic inference engine is a system which will be able to make reasoning about geographic knowledge. Whereas a conventional inference engine is only based on logic, such a system must integrate topology, computational geometry, statistics and spatial analysis because geographic rules necessitate those aspects to be modeled.

The general structure is illustrated in Figure 3.7. The core consists of a geographic inference engine working with the geographic knowledge base together with an input and an output. The input involves the description of a geographic project, such as:

– where is the best place to put a new airport, a new hospital, a new stadium, etc.?

– is this new construction project compliant with planning rules?

– what is the best mode or the best way to get from A to B?

– how do we organize a plan for green spaces in a city?

– how do we reorganize common transportation?

– etc.

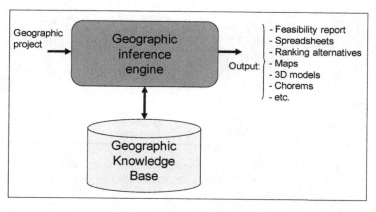

Figure 3.7. *General architecture of a geographic inference system*

As for output, a feasibility report consisting of textual or cartographic issues is more common. Among textual issues, let us mention, outside error or inconsistency reports, what are essentially explanations regarding the possible achievement of the project, and the comparison and ranking of alternatives. With cartographic issues, maps can be good candidates and sometimes, chorems (see Chapter 11 and [DEC 11]) can be an elegant way to summarize the results.

3.3. Towards the integration of external knowledge

In Figure 3.4, the general structure of a geographic knowledge base system was given. Now, it is possible to transform this structure for smart governing and planning. First this structure integrates only what we can call "internal knowledge", or "*intra muros* knowledge" corresponding to the jurisdiction area in contrast with "external knowledge". External geographic knowledge (or *extra muros* knowledge) means knowledge the location of which is outside the jurisdiction: it includes neighboring knowledge located at the vicinity of the jurisdiction (called also crown knowledge) and outside knowledge coming from everywhere else. For instance, outside knowledge can model experiments and good practices in other cities. Figure 3.8 depicts those categories.

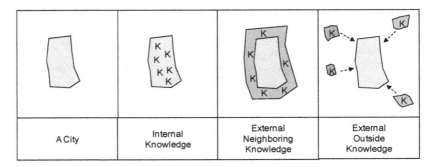

Figure 3.8. *Various categories of geographic knowledge, internal, neighboring (or crown) and outside*

In this book, we will not examine the structure of this kind of knowledge. Let me only say that outside knowledge must be perhaps organized so that case-based reasoning can be performed.

3.4. Prolegomena and principles

Now that the key elements of geographic knowledge are presented, it is time to clarify the context in which geographic knowledge systems could be designed. In order to do so, some principles must be identified. Those principles come from the special nature of geographic data and information. But before outlining those principles, some preliminary concepts called prolegomena must be defined [LAU 14].

3.4.1. *Prolegomena*

These prolegomena will include some preliminary consideration to structure any GKB, that is to establish the foundations so that data can be transformed into geographic knowledge. They are organized as follows:

– the two first prolegomena state the origin of geographic data;

– the next two solve some particular cases of data transformation and advocates for the necessity to deal with consistent data;

– the next two deal with updating of data;

– the next five ones structure objects and geographic information;

– and the last one presents the well-known Tobler's law [TOB 70].

– *Prolegomenon #1* (3D +T objects): "*All existing objects are tridimensional and can have temporal evolution*; lower dimensions (0D, 1D and 2D) are only used for modeling (in databases) and visualization (in cartography)". Unlike geodetic objects which were created by man, all features are 3D, can move, can change their shape and can be destroyed.

– *Prolegomenon #2* (Acquisition by measurements): "*All basic attributes (spatial or non-spatial) are obtained by means of measuring apparatuses having some limited accuracy*". Now more and more data come from sensors; in addition, citizens can be seen as sensors [GOO 07]. In other terms, the word "apparatus" must be taken in a very wide sense, from sensors to census, etc.

Within metadata, accuracy is perhaps one of the most important features, but too often real applications do not care enough about accuracy. One of the big practical difficulties is when different subsets of databases have been acquired with varying accuracy.

– *Prolegomenon #3* (Continuous fields): *"Since it is not possible to store the infinite number of value points in a continuous field, some sampling points will be used to generate the whole field by interpolation"*. As a consequence, special data structures must be developed in conjunction with interpolating functions to estimate value anywhere in the field. See for instance [VCK 95, GOR 01, KAN 02].

– *Prolegomenon #4* (Raster-vector and vector-raster transformations): *"Procedures transforming vector-to-raster data and raster-to-vector data must be implemented with loosing less accuracy as possible"*. Any geographic knowledge system must include those procedures.

– *Prolegomenon #5* (From Popper's falsifiability principle [POP 34]: *"When a new apparatus delivers measures with higher accuracy, these measures supersede the previous ones"*. The practical consequence is that as a new generation of data comes, geographic data and knowledge basis must integrate those data and remove the previous data. But alas, due to the acquisition cost, a lot of actual systems are based on "obsolete" data.

– *Prolegomenon #6* (Permanent updating): *"Since objects are evolving either continuously (sea, continental drift) or event-based (removing building), updating should be done permanently respectively in real-time or as soon as possible"*. Remember that "updating" in computing means three different things, (*i*) a characteristics of an object has varied (for instance, land use in a parcel), (*ii*) the class of an object (so its description) has varied (a building formerly a residence is now for business), (*iii*) an error has been discovered in this object and then corrected (for instance, wrong coordinates or attributes). This prolegomenon implies that any procedure to check or increase data quality must be invoked.

– *Prolegomenon #7* (Geographic metadata): *"All geographic databases or repositories must be accompanied with metadata"*. The necessity to accompany data by information regarding lineage and accuracy was first observed in the GIS domain. More precisely, now the International Standard ISO 19115 "Geographic Information – Metadata" from ISO/TC 211[2] provides information about the identification, the extent, the quality, the spatial and temporal schema, spatial reference, and distribution of digital geographic data. Practically, many geographic databases do not implement the whole standard, but only the more important aspects, because it is very time-consuming. Moreover, metadata must be also updated when necessary.

– *Prolegomenon #8* (Cartographic objects): "*In cartography, it is common to eliminate objects, to displace or to simplify them*". This is due to ensure a maximal readability of maps.

– *Prolegomenon #9* (One storing, several visualizations): "*A good practice should be to store all geographic objects with the highest possible accuracy and to generate other shapes by means of generalization*". This can be seen as an extension of the well-known Douglas-Peucker's family of methods and algorithms for generalization [DOU 73].

– *Prolegomenon #10* (Place names and gazetteers): "*Relationships between places and place names are many-to-many*".

As previously enumerated, Mississippi is the name of a river and the name of a state. The actual city of Rome, Italy, is larger than the same Rome in Romulus's time. The main consequence is that unique feature identifiers must be defined since "popular names" are not so easy to digitally manipulate. Chapter 8 will be devoted to this problem.

– *Prolegomenon #11* (Geographic ontologies): "*All geographic object types are linked to concepts organized into a geographic ontology based on topological relations*". When necessary, raster information can be included into ontologies. For instance, roof textures can be used to identify a building, a wood texture for a wood, a corn field texture to a corn field, possibly with different levels of maturity.

---

2 http://www.iso.org/iso/catalogue_detail?csnumber=26020.

– Prolegomenon #12 [TOB 70]: *"Everything is related to everything else, but near things are more related than distant things"*. This statement may be seen as a key concept also for geographic data mining.

3.4.2. Principles

Now that the prolegomena are defined, principles governing geographic knowledge may be listed in order to get robust reasoning and retrieval. The principles are organized as follows:

– the first three concern the origin of geographic knowledge;

– the next seven ones, the transformation of geographic knowledge;

– the last two ones take the environment into account.

– Principle #1 (Origin of geographic knowledge): *"Spatial knowledge is hidden in geometry whereas geographic knowledge comes also from non-spatial attributes"*. In other words, geographic knowledge is implicit and the question is whether to make it explicit. We can derive from coordinates that New York City is west of Paris, and same kind of relations for all cities throughout the world. The good practice is to derive knowledge on-demand when necessary.

In addition, data coming sensors will support geographic knowledge whereas any indicator will be seen as composite knowledge derived from measures.

Some geographic knowledge can be extracted from data mining techniques.

– Principle #2 (Knowledge cleaning): *"All geographic data, once captured, must be cleaned to remove errors and artifacts to get consistent knowledge"*. This principle is directly connected with Prolegomenon #6 since all automatic acquisition system may include errors or anomalies.

For instance, any airborne laser beam when capturing digital data for terrain or elevation, can intercept a bird: in this case, the captured data will not be the terrain altitude but the bird altitude. Based on this principle, several procedures to cleanse and to increase geographic knowledge quality must be invoked.

However, in practical situations, geographic data or knowledge bases can still encompass some remaining (not yet discovered) errors, thus implying often wrong results in treatment and reasoning. End-users must take care of those errors.

– *Principle #3* (Knowledge enumeration): "*It is not necessary to enumerate all possible chunks of geographic knowledge*". For instance, if one has n objects, then $(n-1)/2$ North–South relationships can also be derived accordingly. Indeed, it is truly possible to derive them automatically when reasoning.

In other words, since any geographic knowledge repository is infinite (intensional), only implicit knowledge is stored, but other knowledge chunks can be derived when necessary.

– *Principle #4* (From geoid to plane): "*On small territories, a planar representation is sufficient whereas for big territories, Earth rotundity must be taken into consideration*". But the question is "how to define a small or a big territory"? A solution can be to define a threshold, for instance a 100 km wide square.

– *Principle #5* (Visualization and visual acuity): "*Cartographic representation is linked to visual acuity*". Here again thresholds must be defined. In classical cartography, as previously said, the limit ranges from 1 mm to 0.1 mm. The same reasoning is valid for cities or small countries such as Andorra, etc. In these cases, the "holes" in Italy or in France disappear cartographically. This point will be detailed in Chapter 4.

For instance, let us consider an A4-format map showing Roman churches in France. These churches must stay in focus when, due to scale, they should disappear.

– *Principle #6* (Crispification): "*At some scales, every fuzzy object becomes crisp*". Section 4.4 will develop this aspect.

– *Principle #7* (Relativity of spatial relations): "*Spatial relation varies according to scale*". Commonly, one says that a road runs along a lake. But in reality, in some place, the road does not run really along the water of the lake due to beaches, buildings, etc. Chapter 5 will develop this aspect.

As a consequence, in reasoning, what is true at one scale can be wrong at another. So, any automatic system must be robust enough to deal with this issue.

– *Principle #8* (Transformation into graph): *"Every set of ribbon[3] or linear objects can be transformed into a graph"*. Indeed, reasoning with graphs is often easier than reasoning with computational geometry. For instance this kind of transformation can be used for roads, rivers, metrolines, sewerages, etc. This problem will be developed in Chapter 4.

– *Principle #9* (From pictorial to geographic objects): *"Any group of pixels having same characteristics located in a satellite image or in an aerial photo can be regrouped into a pictorial object; this pictorial object can be conferred a geographic type and possibly using an ontology"*. Indeed, as soon as a pictorial object is recognized, its type will be identified and can be a part of a geographic object. For instance, a roof texture and an adjacent garden texture can reveal a parcel.

– *Principle #10* (Visualization constraints): *"The spatial relations between objects must hold after generalization"*. In Figure 3.9, an excerpt of the English Riviera coast along the Channel is showed. Suppose we generalize the shoreline by a single line: the city of Eastbourne will be in the middle of the sea whereas Plymouth and Bournemouth will stay in the mainland.

As a consequence of Prolegomena #5 and #6, when better or newer data supersede old data, topological constraints must hold on.

One of the difficulties about this principle is not to follow the constraints, but to ascertain that all visualization constraints are listed. In other words, how do we prove that the list is exhaustive, irredundant and consistent? Here lies a technological barrier.

---

3 The concept of ribbon will be extensively developed in section 4.2.

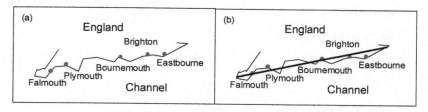

Figure 3.9. *Visualization constraints. a) before generalization; b) generalization of the coastline, but harbors are badly located*

– *Principle #11* (Influence of neighbors): *"In geographic repositories, do not forget that objects at the vicinity (outside the jurisdiction) can have an influence"*.

This is a consequence of Tobler's law (Prolegomenon #12); however the great majority of existing GIS do not follow this law. By using the concept of external knowledge this issue will be dealt with.

– *Principle #12* (Cross-boundary interoperability): *"Any geographic repository must provide key-information to ensure cross-boundary interoperability"*.

This problem has been rapidly presented in section 3.1.5. It can also concern terrain matching at the border. Do not forget that by applying Principle #11, a buffer zone is already integrated into our geographic knowledge base, and some discrepancies can occur. This principle drives the design of consistent distributed geographic knowledge base systems. It will be detailed in section 12.2.

3.5. About quality of geographic knowledge bases

Concerning geographic data quality, several studies were originally made by [GUP 95] and [ANT 91], and so now conventional spatial data quality components are as follows:

– lineage;

– accuracy;

– resolution;

– feature completeness;

– timeliness;

– consistency;

– quality of metadata.

For more details about geographic object quality, I invite the reader to examine Chapter 4 of [LAU 01]. For Geographic Knowledge Bases, the criteria for quality assessment of the other components are more or less the same. However, regarding consistency and completeness, evaluating the consistency of ontologies, gazetteers and rules is a greater challenge. Indeed, it is not enough to discover that two rules are contradictory, but also that two sets of rules are implying contradictory consequences. More studies are needed to define quality for geographic knowledge bases more adequately.

3.6. About multimedia knowledge

All humans can easily recognize faces, landscapes, monuments, music, odors and so on. Often, they have difficulties in describing with words those chunks of knowledge which can enter the category of multimedia. In my previous book [LAU 01], there were extensive descriptions regarding multimedia data engineering for urban and environmental planning.

Now the problem is "*do we have to consider multimedia knowledge?*". It is obvious that multimedia can be used as attribute descriptions such as pictures of buildings; but more generally, what about collections of aerial photos and satellite images and urban noise?

A solution can be to analyze signals, pictures and videos in order to extract some patterns. Those patterns can be the source to encode knowledge about multimedia, but this is not exactly knowledge in multimedia form.

Consider, for instance, the following rules:

– if the bell tolls, I know that death lurks (as Hemingway should have said!);

– if I hear a fire siren, I am informed that a fire is nearby.

Those rules can prefigure multimedia rules. But more generally their encoding and processing is not so easy knowing that basic data are coming from sensors.

Let me quickly note that modeling multimedia geographic knowledge is a new track of research and outside the scope of this book.

3.7. First conclusion on GKS

Where does geographic knowledge come from? Essentially from people and from data as some of them are acquired from sensors. So, we can enrich Figure 1.7 by adding a sort of intermediate layer between smart territory and the physical layer, that is connections, open data and sensors (Figure 3.10). This intermediate layer constitutes the knowledge infrastructure with two components, human intelligence and machine-processable knowledge.

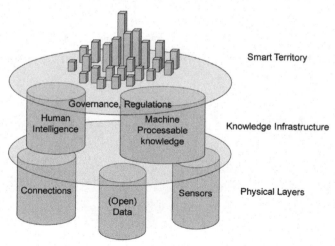

Figure 3.10. *The three layers for Smart Territory governance and regulations*

To summarize what was previously said, a geographic knowledge base, *GKB*, will be formally defined as follows:

$$GKB \equiv (Terr, \lambda, \Omega, GO, \Gamma, REL, \Sigma, RULES, PROJECTS)$$

With $GO \equiv \{O_1, O_2, \cdots O_m: m \in N\}$

in which *Terr* defines a territory, which is a part of *Earth*, λ a language, *GO* is the set of all geographic objects stored in *GKB*, Γ a gazetteer, Ω an ontology, *REL* a set of relationships between geographic objects, Σ a set of structures linking some geographic objects, *RULES* a set of rules and *PROJECTS* a set of old and ongoing projects.

Note that among *GO*, some of them are complex geographic objects and will be examined in Chapter 7. Also, the goal of territorial intelligence projects is often to create new complex objects or modify or replace some elder objects, for instance in urban renewal.

In the subsequent chapters, all these elements will be examined in detail.

4

Geographic Objects

The objective of this chapter is to revisit the modeling of geographic features for knowledge engineering. As previously enumerated, the peculiarity of geographic objects is that they have two facets, geometric and semantic. As by their semantics, that is their types, the can be considered as conventional objects, whereas by their geometric facet they belong to mathematical objects.

In this chapter, both facets will be examined. However, a third facet is important, namely identification (Figure 4.1); this aspect will be treated in Chapter 8. After rapidly dealing with semantic aspects, the core of this chapter will examine mathematical aspects and their consequences.

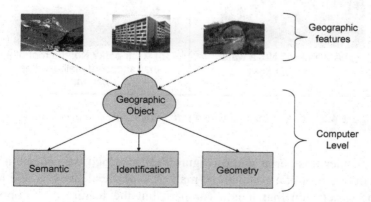

Figure 4.1. *Triple aspects of geographic objects*

4.1. About the semantics of geographic objects

In this section, two aspects must be examined:

– categories or classes to assign a geographic object,

– and geometric shapes.

4.1.1. *Categories or classes*

All humans can distinguish a river from a road, a mountain from a parcel, a city from a lake, etc. However, on facing these features the question is "what is it?". Each language has its own answer, not only by conferring different names, but also because categories are different. For instance, the concept "river" has two translations in French, "*fleuve*" when a river flows to the sea, and the others are only named "*rivière*". Notice that there is a topological relation between "*fleuve*" and sea, and between "fleuve" and "*rivière*", whereas "river" does not bear this kind of connotation. As a consequence, very small rivers flowing to the sea are classified into the French category "*fleuves côtiers*". This problem will be explored extensively in Chapter 8.

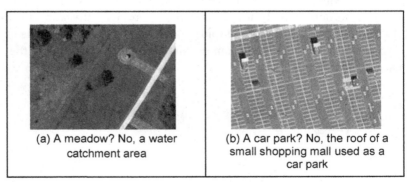

| (a) A meadow? No, a water catchment area | (b) A car park? No, the roof of a small shopping mall used as a car park |

Figure 4.2. *Some problems linked with geographic object classification*

Consider this aerial photo (Figure 4.2) in which we can see a car park on the roof of a commercial mall. A superficial analysis will lead to the conclusion that it is a car park but the feature of "shopping mall" is perhaps more important. In addition, in the same aerial photo, we can see a meadow, but in reality this a catchment zone for the nearby water pumping station. In other words, sometimes it is not so easy to confer a unique category or class to a real feature.

4.1.2. Place names and identifiers

A lot of geographic features have names and in some cases, several names. As such, the same name could be assigned to various different features (see Prolegomenon #10). To solve this problem, computer identifiers were created (IDs). As this solution allows us to easily distinguish two computer objects, for humans this is not very convenient. Moreover, in different databases, the same features can have different identifiers. Look for example for the European Rhine River: there is no apparent reason that it has the same computer ID in Germany and in Romania. The role of gazetteers is to solve this problem. Initially created as place name dictionaries, they are now databases or more special knowledge bases. Chapter 8 will explore with this aspect.

However, this chapter will focus essentially on the mathematical modeling of geographic objects.

4.1.3. Geometric types

Our world is tridimensional (see Prolegomenon #1) with moving objects (x, y, z, t). Usually, in cartography, only x and y are used in the so-called 2D model. When the z dimension is used, the model is called 3D. Sometimes, for instance for terrain modeling, the third dimension is taken as an attribute of a 2D point (x, y); in this case, the model is usually called 2.5D.

In this book, overall 2D objects will be considered. But sometimes when time and elevation will be taken into account, it will be stated.

Among geographic objects, two categories can be distinguished, those which have known boundaries (parcels, roads, etc.) and those with undetermined boundaries (mountains, marshes, deserts, etc.).

For those with known boundaries, conventional geometry will be used and objects will be considered as precise. Whereas when undetermined boundaries, models deriving from fuzzy sets will be used.

However, one of the problems comes from the curvature of the Earth. As planar geometry could be easily used for smaller territories, for bigger objects this rotundity must be taken into account. In other words, the coordinates x and y must be interpreted as latitude and longitude (according to the WGS 84 system), and z as sea level or altitude.

For centuries in cartography, scales were used as a key-element. However, in geoprocessing, scale is only a parameter for visualization and not for storage. As previously said, the concepts of granularity of interest and visual acuity can be invoked.

Do not forget that feature shapes are always simplified overall on smaller scales: in order to increase readability, lines are generalized that is some points are removed (see Prolegomenon #8) thanks for instance to the well-known algorithm designed by Douglas-Peucker [DOU 73]. In addition, depending on the context sometimes some cartographic objects must be enlarged or slightly moved.

Again, for decades, geometric models of geographic features have been based on points (0D), lines (1D), areas (2D) and volumes (3D). However, points and lines do not exist in nature since all objects are 3D and moving. Geography for its part is mainly 2D. In the majority of GIS (geographic information systems) software products, rivers and roads are modeled as lines, sometimes with a width, which is strange from a mathematical point of view. To solve this problem, the concept of ribbon will be introduced.

4.2. From lines to ribbons

Since lines do not exist in the real world, except perhaps lines such as the Equator, the meridians and the parallels, in a recent paper [LAU 14], I have proposed to use ribbons to model what it is common to call linear objects such as roads and rivers. A ribbon can be defined as a line with a width.

It is because of these curves (circle portions, clothoids) that roads are not rectilinear. Therefore the idea of modeling lanes by rectangles is insufficient. In order to deal with this important characteristic, a more general definition is needed. From a mathematical point of view, a ribbon can be defined as a transformation of a long rectangle. Figure 4.3 illustrates this principle. Let's call a ribbon R and ρ a rectangle of length l and width w. One can state $R = H$ (ρ), in which H is a taenic transformation[1].

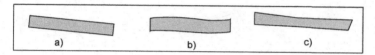

Figure 4.3. *Various types of ribbons:*
a) rectangular ribbon; b) ribbon; c) loose ribbon

4.2.1. Rectangular ribbons

A rectangular ribbon R is a long rectangle. Let us call the ends the two smaller extremities and the sides larger ones. The ribbon length ratio rl ($rl= l/w$) is supposed to be much greater than a positive value rL; a possible minimum value is 5 ($rl>rL\geq 5$).

Let us call the medium line between two ends located at distance $h =w/2$ from the sides of the rectangle. Let us note $Skel(R)$, $End1(R)$, $End2(R)$, $Side1(R)$, $Side2(R)$, respectively, the skeleton of R, its two ends and its two sides.

4.2.2. Ribbons and taenic transformation

A taenic transformation (Figure 4.4) transforms a 1D curve into a 2D area with the following properties. Let us note $y=f(x)$ a curve which is supposed to be continuous and differentiable. In each point of the curve, let us consider two points U_1 and U_2 at a distance h from the

1 From Ταινία, ancient Greek for ribbon.

curve located at the perpendicular of the first derivative. Let's call them corresponding points. The loci of those corresponding points form two curves C_1 and C_2 respectively. Knowing that $dy / dx = tg(\alpha)$, we can write respectively:

So for C_1: $\begin{cases} x_1 = x + h \times \sin(\alpha) \\ y_1 = y - h \times \cos(\alpha) \end{cases}$ and for C_2: $\begin{cases} x_2 = x - h \times \sin(\alpha) \\ y_2 = y + h \times \cos(\alpha) \end{cases}$.

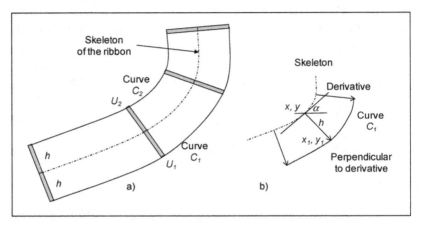

Figure 4.4. *Construction of a regular ribbon with a taenic transformation: a) regular ribbon; b) details of the construction of a regular ribbon*

Consider the derivative at corresponding point U_1, one has:

$$\frac{dy_1}{dx_1} = \frac{dy_1}{d\alpha} \times \frac{d\alpha}{dx_1} \text{ but } \frac{dy_1}{d\alpha} = -h \times (-\sin(\alpha)) = h \times \sin(\alpha)$$

and $\dfrac{dx_1}{d\alpha} = h \times \cos(\alpha)$.

So $\dfrac{dy_1}{dx_1} = \dfrac{dy_1}{d\alpha} \times \dfrac{d\alpha}{dx_1} = \dfrac{h \times \sin(\alpha)}{h \times \cos(\alpha)} = tg(\alpha) = \dfrac{dy}{dx}$.

A similar computation can be done for U_2 giving $\dfrac{dy_2}{dx_2} = \dfrac{dy}{dx}$. So the consequence is that the derivatives of the corresponding points are equal to the derivative at the skeleton and the angle is the same.

Now, consider a ribbon end. We can write the following dot product (*DP*) between the tangent in U_1 and the concerned end:

$$DP = (h \times \sin(\alpha) \times \Delta x_1 + (-h \times \cos(\alpha) \times \Delta y_1)).$$

So if $\Delta x_1 \to 0$, we can see that the dot product tends also towards zero which means that the angle is right. As a conclusion, both ends are orthogonal to the sides.

4.2.3. *Loose ribbons*

Starting from any polygon P, what are the conditions to consider it as a loose ribbon? Indeed often due to measurement errors (for instance roads) and other reasons (for instance rivers), ribbons are not always perfect rectangles. For solving this problem, let us consider its equivalent rectangle.

The first step is to consider all vertices of P (Figure 4.5(a)) and by the least square method to compute the regression line $y=mx+q$ (Figure 4.5(b)). Let us define the angle so that $tg(\theta)=m$. Then we make a rotation of $-\theta$ so that the regression line is parallel to the x-axis (Figure 4.5(c)). Then, we sort all vertices according the ascending values of respectively x and y coordinates. We determine minimum and maximum according to those orders. Along x, the mid values of the two first and the two last will determine ribbon ends; and along y, the mid values of the two first and the two last will determine sides (Figure 4.5(d)); those value will determine the equivalent rectangle of P noted $ER(P)$.

Now, let us compare the areas of P and $ER(P)$. Generally speaking, there is a small discrepancy between those values. A solution is to lightly modify l and w to reach the exact value. Let us note

A_1=*Area*(P) and A_2=*Area*($ER(P)$). Generally speaking, they are not equal. To get them equal:

$$A_2 = (w + \Delta w) \times (l + \Delta l) = w \times l \times (1 + \frac{\Delta w}{w}) \times (1 + \frac{\Delta l}{l}) = A_1 \times (1 + \frac{\Delta w}{w}) \times (1 + \frac{\Delta l}{l})$$

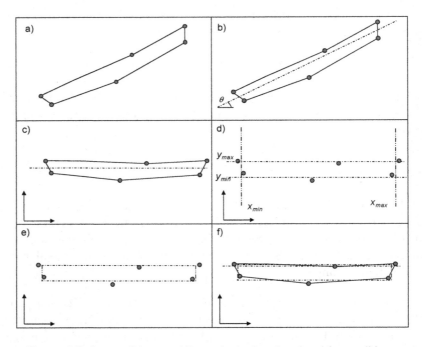

Figure 4.5. *Loose ribbon and its equivalent rectangle: a) loose ribbon; b) the regression line; c) rotation; d) determination of the two minima and the two maxima; e) equivalent rectangle; f) loose ribbon and its equivalent ribbon before the back rotation*

Let us suppose that we want to impose modification in the same proportion. So we can write $\dfrac{\Delta w}{w} = \dfrac{\Delta l}{l} = t$ the modification ratio, so giving $A_2 = A_1 \times (1 + t)^2$. Its value is $t = \sqrt{\dfrac{A_2}{A_1}} - 1$, or $t = \dfrac{\sqrt{A_2} - \sqrt{A_1}}{\sqrt{A_1}}$. With this modification, both areas will be equal.

If the length ratio *rl* of the equivalent rectangle is greater than the threshold value *rL* then the polygon *P* is considered as a loose ribbon *R* =ERR(*P*).

In order to simplify this presentation, we will now only consider ribbons as rectangles, that is, the *H* transformation becomes the identity transformation $R = H(\rho)$ so $R = H(\rho) = \rho$.

In Figure 4.6, an example of a road modeled by ribbons is presented in which one can distinguish several ribbons, namely for bus lanes, bike lanes, median and so on. Immediately we that ribbons can have specific relations between them.

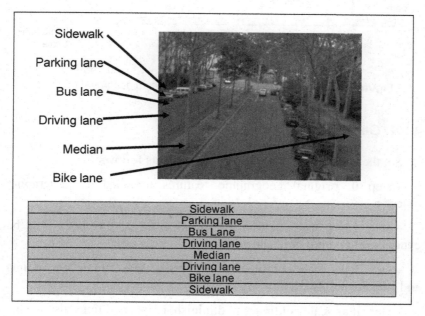

Figure 4.6. *An urban road divided into several ribbons*

4.3. Mutation of object geometric types

According to the concept of granularity of interest, geographic objects can mutate according to rules which are given below (see examples Figure 4.7). As scale diminishes, an area will mutate into a point and then will disappear, and a ribbon will mutate into a line and then will disappear.

100 Km wide city	Invisible	Reduced to point			Visible area	
1 ha wide hamlet	Invisible		Reduced to point		Visible area	
100 m wide motorway	Invisible		Reduced to line		Visible ribbon	
1 m wide path	Invisible			Reduced to line	Visible ribbon	
	10^{-10}	10^{-8}	10^{-6}	10^{-4}	10^{-2}	1 Logarithmic Scale

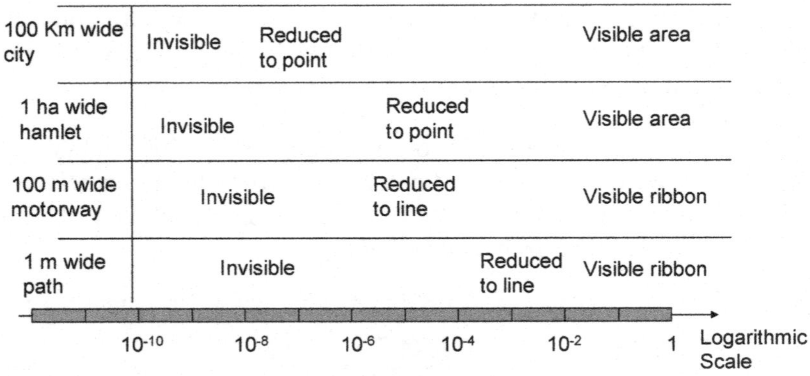

Figure 4.7. *Generalization of geographic objects according to scale*

4.3.1. *General process (GRD process)*

So, the complete process can be modeled as follows:

– step 0: original geographic features modeled as geographic objects;

– step 1: as scale diminishes, small areas and ribbons will be generalized and possibly can coalesce;

– step 2: as scale continues to diminish, areas mutate into points and ribbons into lines;

– step 3: as scale continues to diminish, points and lines disappear.

Let us call this process "generalization-reduction-disappearance" (GRD process).

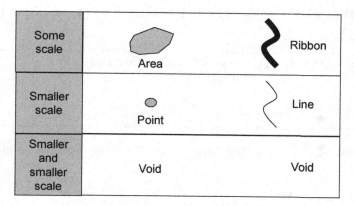

Figure 4.8. *Mutation of geographic objects*

4.3.2. *Rules of visual acuity applied to geographic objects*

With the thresholds ε_i, ε_{lp} previously defined, we can formally obtain (in which *2Dmap* is a function transforming a geographic object to some other scale possibly with generalization) the following rules:

a) **Rule #4.1:** Disappearance of a geographic object (O) at scale σ noted O^σ:

$\forall O \in GO, \forall \sigma \in Scale,$ G-$Type(O) = Area, O^\sigma \equiv 2Dmap(O, \sigma)$: $$Area\,(O^\sigma) < (\varepsilon_{lp})^2$$ $$\Rightarrow$$ $$O^\sigma = \varnothing.$$	Rule 4.1

b) **Rule #4.2:** Transformation of an area into a point (for instance the centroid of the concerned object, for example defined as the center of the minimum bounding rectangle, see section 5.1.2. for details):

$\forall O \in GO, \forall \sigma \in Scale,$ G-$Type(O) = Area, O^\sigma \equiv 2Dmap(O, \sigma)$: $$(\varepsilon_i)^2 < Area\,(O_\sigma) < (\varepsilon_{lp})^2$$ $$\Rightarrow$$ $$\{G\text{-}Type(O^\sigma) = Point; O^\sigma = Centroid(O)\}$$	Rule 4.2

c) **Rule #4.3**: Transformation of a ribbon R into a line (for instance its axis):

$\forall O \in GO$, G-$Type(O)$=$Ribbon$, $\forall \sigma \in Scale$, $O^\sigma \equiv 2Dmap(O, \sigma)$: $\varepsilon_i < Width(\ O^\sigma) < \varepsilon_{lp}$ $\Rightarrow$ $\{G$-$Type(O^\sigma) = Line;\ O^\sigma = Axis(O)\}$	Rule 4.3

In the sequel, those rules will be called geographic object mutation rules.

4.4. Fuzzy geographic objects

Usually, two categories of geographic object description can be distinguished – crisp and fuzzy. Crisp objects must have well-defined boundaries such as administrative objects (countries, regions, provinces, natural parks, parcels, etc.) and manmade objects such as streets, buildings.

Other objects, for instance natural features can be defined as *crisp objects*; but there are difficulties. A river by some scales can be defined as a line whereas sometimes the expressions such as a minor or major bed are used. Even some dry rivers can be without water. For the seas, according to the tide levels, geometric shapes can be different. One of the more salient examples is "Mont Saint-Michel", in France, which is roughly only 1 km^2 wide at high tide and several squared kilometer wide at low tide.

Where does a hill begin, What is the upper limit of a valley? Where does begin a marsh? Those are common questions in which features can be modeled as fuzzy geographic sets.

For those objects, fuzzy sets can be used in which some membership grades can be defined (Figure 4.9) [ZAD 95]. An interesting model [COH 96] is the egg-yolk model with two parts, the core (the yellow part) and the extension, the white part of the egg. For instance for a river, the "yolk" represents the minor bed whereas the "egg" modeled its major bed. Another example is given in Figure 4.10 in which the mangrove and the jungle are modeled with the egg-yolk representation.

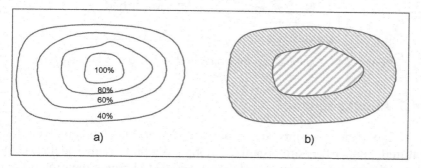

Figure 4.9. *Fuzzy geographic object: a) different membership grades; b) the egg-yolk representation*

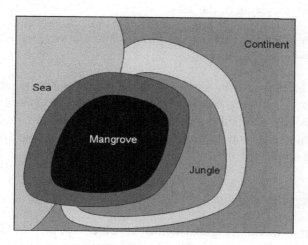

Figure 4.10. *Fuzzy geographic features*

Again, the egg-yolk model can be used to model ribbons: so a fuzzy ribbon can have a wider ribbon, including a more narrow ribbon. This model can be applied to modeling rivers, each of them with their minor bed and major bed exemplified in Figure 4.11.

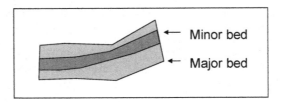

Figure 4.11. *Example of a river modeled by an egg-yolk-style ribbon*

4.5. About altitude

Two main concepts are important concerning data involved for attitude: either they are organized along a digital terrain model (DTM) or a digital elevation model (DEM). As a DTM stores data regarding the ground or natural terrain, DEM considers objects above the ground. Moreover, the same methods can be used to store points representing the depth of the seas.

However, before analyzing operations and relations, let us consider a few issues regarding modeling.

Although they can be represented by sets of points together with altitudes (x, y, z), three families of models (Figure 4.12) exist namely irregular points, gridded points and contour level curves. In addition, there are methods to pass from one model to another, but sometimes sacrificing quality.

4.5.1. *Irregular points*

The first model (Figure 4.12a) considers points as randomly distributed over the terrain. Then, a Delaunay triangulation is run to form triangles. Often this model is called Triangulated Irregular Network (TIN). With the relational database model, several methods are possible, for example based on three relations:

```
TRIANGLE   (Triangle_ID,   Point1_ID,   Point2_ID,
Point3_ID)
POINT (Point_ID, x, y, z)
```

4.5.2. *Gridded points*

This model (Figure 4.12(b)) is based on a regular squared grid, for instance every 100 m. For storing, a matrix can be used.

One of the difficulties of this model is that sometimes the point supposed to be on the grid is covered by a building. In this case, an approximation must be used to reconstitute the natural terrain.

4.5.3. *Contour level curves*

The third model (Figure 4.12(c)) is not used for storing but for visualization. Indeed, this method of visualizing altitude is very old and common in cartography. Usually a step (for instance 10 m) is used.

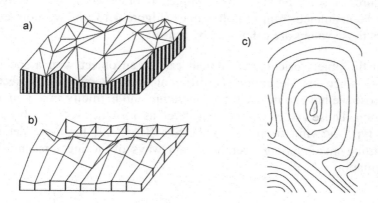

Figure 4.12. *Examples of terrain models: a) based on triangles (TIN); b) squared grid; c) contour levels*

4.6. Continuous fields

Consider continuous phenomena such as temperature, pressure, wind, air pollution or noise: they can be described by continuous fields. They can be defined by a function which is valid everywhere $f(x, y, z, t)$ on a geographic domain (2D, 3D or even 3D+T); it can be a scalar function, for instance for instance for temperature or a vector function for winds or noise levels. Let us note that the modeling of terrains and elevation can also be based on continuous field theory. Formally speaking [LAU 00], a 2D field is defined by:

$F = (D, P, V, f)$ where:

$- D$ is the location domain of the field that is the region it represents $D \subseteq R^2$;

$- P$ is the set of positions where the values of the phenomenon are known, $P \subseteq R^2$;

$- V$ is field domain value;

$- f : D \to V$ is an estimation function describing the phenomenon.

Practically speaking, we do not know the function f, but we have measures on some points, called samples. In other words, this function must be estimated from sample values (Figure 4.13). Now, more and more, those values are calculated by sensors.

In the object orientation, a field can be an object with a special status, since its definition is different from geographic objects. Suppose we have to define the temperature and humidity in a point or a region. Those values will be considered as attributes with an abstract data type deriving from the fields. See [LAU 00] for more details regarding manipulating operations and ways of storing this kind of information.

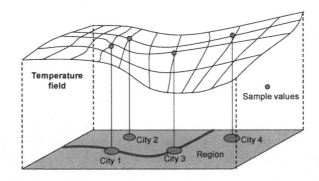

Figure 4.13. *Example of a continuous field and a few sample values*

Remember that a continuous field is governed by Laplace equation:

$$\frac{\partial^2 f}{\partial x^2} + \frac{\partial^2 f}{\partial y^2} + \frac{\partial^2 f}{\partial z^2} = 0.$$

Since f is unknown and since we have only a few values of this function, the previous model for altitude could be used to model those values in the 2D case. From a knowledge point of view, fields can be modeled as geographic objects, for example, soil temperature in a territory, but with a special characteristic: therefore temperature of a place can be defined as an abstract data type whose value derived from the field [GOR 01].

In this book, the scope is not to develop fields from a mathematical and computing point of view, but rather to analyze them in terms of knowledge engineering. In other words, data coming from fields must be transformed first into objects, and then into knowledge. For instance by examining pressure, we can estimate low-pressure and high-pressure areas and then define cyclones and anticyclones; then we can create rules to "generate" winds.

4.7. Quality and geometric homology relations

Quality is a key-element for any kind of action. Concerning geographic data coming from measures, this problem is crucial because two apparatuses can give different measurements. In my previous books, some chapters were dedicated to this issue, namely ([LAU 93] Chapter 15) and ([LAU 01] Chapter 4), in which spatial accuracy is one of the main components of quality assessment. The metrics used depend on the dimensionality of the entities under consideration. For points, accuracy is defined in terms of the distance between the encoded location and "actual" location. Error can be defined in various dimensions, x, y, z. Regarding points, often spatial accuracy can be presented in terms of a spatial tolerance such as ± 1 foot. For lines and areas, the situation is more complex. This is because error is a mixture of positional error (error in locating well-defined points along the line) and generalization error (error in the points selected to represent the line). The ε band [BLA 83] is usually used to define a zone of uncertainty around the encoded line, within which the "actual" line exists with some probability. See Figure 4.14.

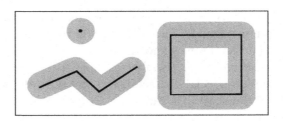

Figure 4.14. *Examples of error band for a point, a line and a polygon*

4.7.1. *Quality control based on rules*

Numerous works have been made [SER 00, BOR 02, BRA 12] and [DEL 15] on geographic data quality control, whether the accuracy of the data, to check the consistency polygons and tessellations (see section 7.3), of the elimination of sliver polygons, or the elimination of the outliers, etc. For example the paper [SER 00] proposes to model topological constraints by the following formalism:

```
Constraint    =    (EntityClass1,    relation,
EntityClass2, Specification)
```

in which: Specification can only have the following values, Forbidden, At least *n* times, At most *n* times, Exactly *n* times.

4.7.2. *Geometric homology*

As previously enumerated (section 3.1.3.), when we want to compare two geographic objects, we need to ascertain whether the geometric shapes (taking measurement errors into account) are similar or not. For that purpose, the ideal will be to create an equivalence relation, that is to say, reflexive, symmetric and transitive. However, due to the previous remarks, one can mention that transitivity is not verified every time. Therefore, let us consider geometric homology relations (denoted as ♊$_G$).

For sharp areal geographic objects (A and B), the boundaries are well known and agreed upon, but there are practically always measurement discrepancies. In this case, to match them, we can compare their geometric shapes (Figure 3.2) and their locations, for instance, by their centroids (section 5.1.12). By definition, two areal geographic objects, A and B, are considered as geometrically homologous iff:

$$Geom(A) ♊_G Geom(B)$$

$$\Leftrightarrow$$

$$\left(\frac{2 \times Area(A \otimes B)}{(Area(A \cup B) + Area(A \cap B))} \leq \varepsilon_1 \right) \wedge Dist(Centroid(A), Centroid(B)) < \varepsilon_2)$$

in which ε_1 and ε_2 are some predefined thresholds. Moreover, in this expression, remember that the symbol $\otimes$ is called "symmetric difference" and is defined as follows $A \otimes B = (A \cap \neg B) \cup (\neg A \cap B)$.

Two remarks can be made. The first one is that this expression can be decomposed into two rules, and second, that this expression can also be valid for ribbons.

When it is to compare islands, we can apply the previous method for matching them. However for other natural objects, this is more complex because sometimes, boundaries are indeterminate, especially for mountains and deserts. For instance, comparing two different geometric representations of the Rocky Mountains based on geometry is not so easy, because two experts can give two different boundaries of the mountains.

For lines, it is a little bit different. Considering two polylines A and B, what is the distance between them? An interesting definition is given by the Fréchet distance which corresponds to the minimum leash between a dog and its owner, the dog walking on a line, and the owner in the other line as they walk without backtracking along their respective curves from one endpoint to the other. The definition is symmetric with respect to the two curves. By noting a, a point of A, and b of B, the Fréchet Distance *Frechet* is given as follows in which *dist* is the conventional Euclidean distance [ALT 95]:

$$Frechet(A, B) = \underset{a \in A}{Max}(\underset{b \in B}{Min}(dist(a, b)))$$

Therefore the homology between two lines A and B will be defined as follows:

$$Geom(A) \bowtie_G Geom(B)$$

$$\Leftrightarrow$$

$$\left(\frac{2 \times Frechet(A, B)}{(Length(A) + Length(B))} \leq \varepsilon_3 \right)$$

4.8. Geographic objects and projects

As previously mentioned, the scope of a geographic inference engine is to reason about new projects. From a formal point of view, a project is essentially a geographic object maybe a complex object (for instance an airport, a bridge, a city-block to be renovated) which does not yet exist; and once the project is approved, this object will be constructed. Finally, it will correspond either to a new feature or a renovated feature, that is with different characteristics. In this book, they will be called projected objects.

For any project, it could of value to store its parameters and its overall evaluation. Since several stakeholders must be considered and several criteria must be invoked, each project will be accompanied by a multi-criterion and multi-actor matrix.

4.9. Final remarks concerning geographic objects

This chapter was devoted to the description of geographic objects which will be sometimes called Geo-objects. From a mathematical point of view, a set *GO* will be constructed. However, in this book, continuous fields will not be considered.

$$GO \equiv \{O_1,\ O_2,\ \cdots\ O_n : n \in N\}.$$

Hence, any geographic object will have:

– an ID named *GeoID* which will be an identifier only used for storing;

– a geographic type as explained in this chapter;

– a geometric shape (the more recent and the most accurate possible), when necessary other less accurate representations will be derived quickly by using generalization algorithms (*Geom*);

– zero, one or several toponyms (see Chapter 8 for details).

Mathematically, they will be modeled as (concerning toponyms, only the more commonly used will be included);

$$O_i \equiv (GeoID_i,\ G\text{-}Type_i,\ Topo_i,\ Geom_i,\ \Omega\text{-}Type,\ (Attribute,\ Value)^*)$$

From a geometric point of view, the type will be as follows {*Point, Line, Area, Ribbon, Void, Null*}.*Void* (also noted $\varnothing$) will be used when the geometry does not exist, and *Null* when the geometry exists but is not known. In this list, we can add *Field* and modifiers such as

– *Sharp* and *Fuzzy* for points, lines, ribbons and areas;

– *Oriented* or *Not Oriented* for lines and ribbons.

So we can write for instance:

G-*Type* ("Lake_Tahoe") = (*Crisp Area*)

G-*Type* ("Rocky_Mountains") = (*Fuzzy Area*).

To simplify notations, let us propose that *Crisp* and *Not-Oriented* are given *by default*.

Regarding the couples (*Attribute, Value*), they will be defined according the type (Ω-*Type*) as given in the ontology, and *Values* can be stored as alphanumeric or multimedia data.

In addition,

– those objects will be linked by spatial or geographic relations (Chapter 5) or even by structures (Chapter 7); moreover other geometric types will be added for describing complex objects;

– they will adhere to rules (Chapter 10);

– and they will be the base for projected objects.

5

Geographic Relations

As the goal of the previous chapter was to examine geographic objects both from geometric and semantic points of view, this chapter will examine the relations they share between them. Topological relations are among the well-known relations. But beyond those relations, many others can connect various geographic objects, for example two twin cities. What is the difference between a spatial relation and a geographic relation? When one says Africa is South of Europe, it is overall a geographic relation though it may also be described as a spatial relation linking those two continents. Even as spatial relations are commonly used in geography, they can be used also in other domains such as robotics, medical imagery, etc. However, there are no clear-cut differences between spatial and geographic relations; maybe we can say that spatial relations are seen more abstract whereas geographic relations are grounded in the Earth. Finally, a geographic relation links two objects located in the Earth.

In addition, within cities for instance, some urban objects may present regular patterns, so giving new forms of spatial relations through shape grammars. Shape grammars are a way to describe configurations of spatial objects presenting regular patterns. In a way, they are another form of geographic relations not linking two objects, but several geographic objects. This aspect will be covered in Chapter 7.

Another important aspect is that geographic relations can vary according to scale. Indeed as geometric types can mutate according to scale, the same phenomenon occurs for geographic relations.

However before exploring those relations, it is important to say a few words on spatial operations because they are the background of spatial relations. Then spatial relations and especially topological relations will be presented. Then several other categories of geographic relations will be detailed.

5.1. Spatial operations

There are a lot of spatial or geometric operations which can be used in geoprocessing. Only a few will be examined in this text, that is minimum bounding rectangles, centroids, buffer zones and convex hulls. Remember that some spatial operations include sophisticated computational geometry algorithms. For more details, refer to [PRE 85].

5.1.1. *Minimum bounding rectangle*

For several more sophisticated operations, it is interesting to determine the smaller rectangle encompassing an area, for instance a polygon, the sides of which are parallel to the coordinate axes. In the list of polygon coordinates, we can easily extract x_{min}, x_{max}, y_{min} and y_{max}. Figure 5.1(a) shows a polygon A, and Figure 5.1(b) its minimum bounding rectangle (*MBR*). In reality, due to the curvature of the Earth, the minimum/ maximum longitude and latitudes, instead of lines, this MBR is limited by part of ellipsoids (Figure 5.1(c)). Let us encode it *MBR(A)*.

5.1.2. *Centroid*

Sometimes, it is interesting to exhibit a point representing an area: let us call in a centroid located in the center of the area. Centroids can be loosely defined as a point in the center of an area. However, mathematically speaking, several possibilities [LAU 93] can be used to define a centroid, for instance (Figure 5.2):

– the barycenter of vertices (Figure 5.2(a));

– the center of gravity (Figure 5.2(b));

– the center of the minimum bounding rectangle (Figure 5.2(c)).

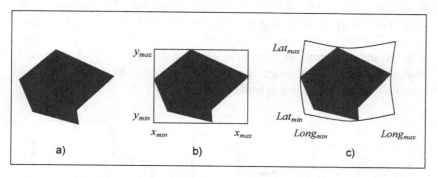

Figure 5.1. *Minimum bounding rectangle: a) initial polygon;
b) its MBR; c) an MBR taking Earth rotundity into account*

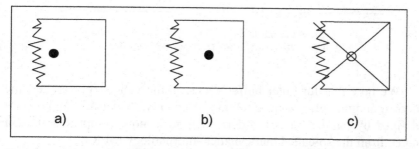

Figure 5.2. *Several definitions of a centroid. a) as barycenter of coordinates;
b) as center of gravity; c) as center of the minimum bounding rectangle*

Generally the last one is used because it is the easier to calculate.
Let us encode it *Centroid (A)*. When the polygon is not connected, the
center of the MBR of the larger component is taken; for instance, for
the US, only the *MBR* of the conterminous States is usually used.
Sometimes the centroid can be outside the territory; consider for
instance Florida or Croatia.

5.1.3. *Buffer zones*

When one wants to get the people living within less than 10 km from the borders of a country, a buffer zone must be defined. More generally, when considering a polygon (Figure 5.3(a)) and a distance *d*, two buffer zones can be defined, an inner buffer zone (Figure 5.3(b)) and an outer buffer zone (Figure 5.3(c)). Let us encode it *Buffer(A,d)* in which when *d* is positive, this is an outer zone and when *d* is negative an inner buffer zone.

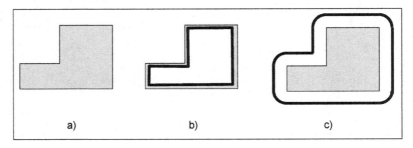

Figure 5.3. *Definition of a buffer zone: a) initial polygon; b) inner buffer zone; c) outer buffer zone*

We note that for outer buffer zones, in the vicinity of vertices, the corresponding buffer zone is rounded. In reality, the buffers are not only defined by parallel lines but derived from a more complex definition taken from the so-called mathematical morphology [SER 82].

5.1.4. *Union, intersection and difference*

One of the major needs and challenging problems in spatial information is to compute the difference, union and the intersection of polygons. For example, considering two polygons *A* and *B* (Figure 5.4(a)), we need to find their union; that is, their joint extent (Figure 5.4(b)) and their intersection – their common area (Figure 5.4(c)). As in their book, Preparata and Shamos [PRE 86] present several union and intersection algorithms, we present only the simplest possible way, based on the slab technique. Each polygon is divided into parallel slabs, usually for the convenience of parallelism with the coordinate axis, as in Figure 5.5, created by drawing lines through the polygon. This procedure creates trapezoids which are easier to compare. Figure 5.6 gives some varied examples to illustrate this method. When they are not intersecting inside the slabs, the comparison is straightforward; a solution is to split those slabs into other smaller slabs. To get the ultimate result, we have to glue all resulting slabs into a polygon, possibly non-connected.

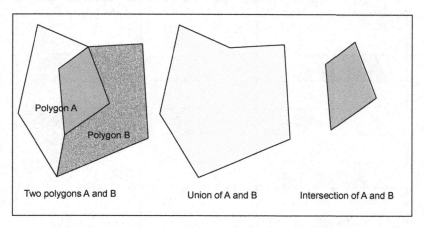

Figure 5.4. *Union of intersections of two polygons*

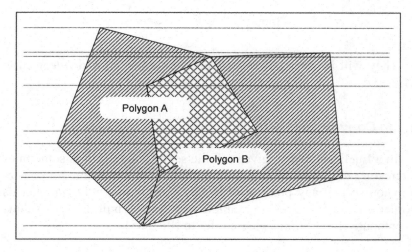

Figure 5.5. *Splitting polygons with parallel slabs*

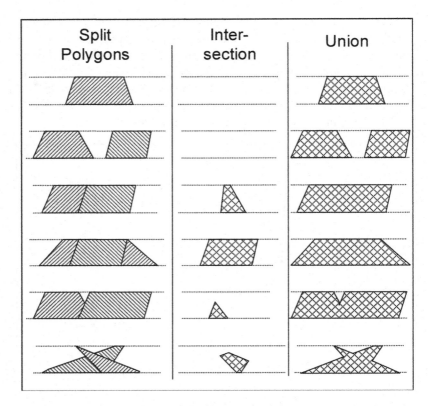

Figure 5.6. *Comparing slabs*

They will be denoted respectively *Union* (A, B) and *Intersection* (*A*, *B*).

5.1.5. *Convex hull*

In a Euclidean plane, given a finite set of points *Q*, it is sometimes interesting to determine its convex hull, namely the minimum convex polygon so that any point of *Q* is either inside this polygon or at its border. Figure 5.7 gives an example of a convex hull. For algorithms to compute convex hulls, please refer to [PRE 85].

Extensions can be found in 3D and higher dimensions. Also, it is possible to determine convex hull for areas. In this book, we will only consider the 2D case, by writing $CH(K)$ in which K will stand either for a set of points or for a set of areas.

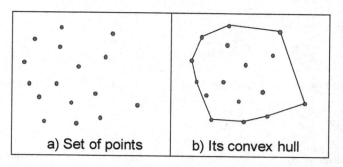

Figure 5.7. *A set of points and its convex hull*

5.2. Spatial relations

First, remember Principle #1 (in section 3.3.1) that spatial relations are hidden in coordinates. In this section, among planar spatial relations, topological and projective relations will be examined.

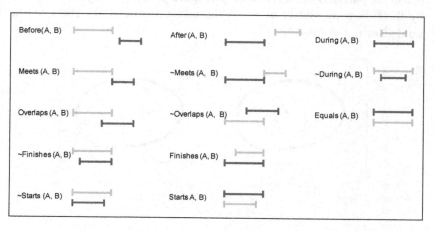

Figure 5.8. *Allen interval topological relations at 1D [ALL 83]*

5.2.1. *Topological relations*

Topological relations such as at 1D, interval Allen relations [ALL 83] (Figure 5.8) and at 2D Egenhofer relations [EGE 91, EGE 94] are well known (see Figure 5.9).

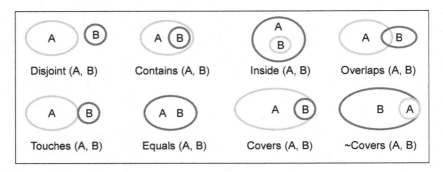

Figure 5.9. *Egenhofer topological relations at 2D [EGE 94]*

To determine the topological relation between two areas, one solution [EGE 92, CLE 93] is to compute the so-called nine intersections. Considering a polygon, let us note $A°$ the inner part, $\neg A$ the outer part and ∂A its boundary (Figure 5.10). The answer is given by the following matrix in which the result of one intersection can be void $\varnothing$ or not void $\neg\varnothing$.

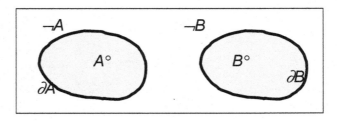

Figure 5.10. *Determining the topological relation between two objects. A° and B° represent the inner parts, ∂A and ∂B the boundaries and ¬A and ¬B the outer parts*

$$R(A,B) = \begin{pmatrix} A° \cap B° & A° \cap \partial B & A° \cap \neg B° \\ \partial A \cap B° & \partial A \cap \partial B & \partial A \cap \neg B \\ \neg A \cap B° & \neg A \cap \partial B & \neg A \cap \neg B \end{pmatrix}$$

For instance for *Touches*, the result is as follows:

$$Touches(A,B) = \begin{pmatrix} \varnothing & \varnothing & \neg\varnothing \\ \varnothing & \neg\varnothing & \neg\varnothing \\ \neg\varnothing & \neg\varnothing & \neg\varnothing \end{pmatrix}$$

In [CLE 93], the authors have developed this matrix by integrating the dimensions of the intersections (0D, 1D or 2D).

With such topological relations, one can easily define relations for geographic features. For instance Figure 5.11 presents a *Touches* relation between a river and the sea.

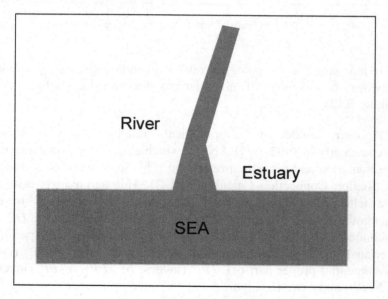

Figure 5.11. *There is a* Touches *topological relation between river and sea, corresponding to the estuary*

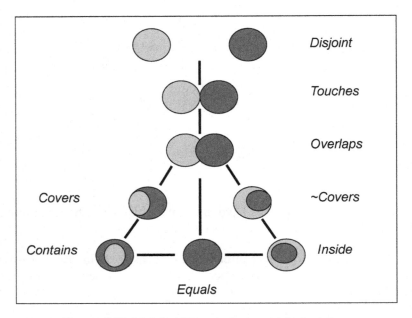

Figure 5.12. *Vicinity of Egenhofer topological relations*

In any case, these topological relations can be organized in a graph showing the vicinity of all relations for the Egenhofer model (Figure 5.12).

Another model for topological relations was proposed independently in 1992, by [RAN 92] which allowed qualitative spatial representation and consistent reasoning. This logic received the name of "Region Connection Calculus" (RCC). This acronym is also the first letter of authors' names. This model is equivalent to the Egenhofer model. The eight relations have different names: *DC* (is disconnected from), *EC* (is externally connected with), *PO* (partially overlaps), *TPP* (is a tangential proper part of), *NTPP* (is a nontangential proper part of), TPP_i (inverse of *TPP*), $NTPP_i$ (inverse of *NTPP*) and *EQUAL* (Figure 5.13).

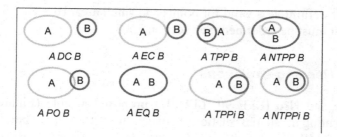

Figure 5.13. *The RCC topological model*

Some remarks must be made concerning the relationships between topological relations and set-theoretical relations. Indeed consider the set *GO* for geographic objects and *Earth* representing the terrestrial globe. As it is clear that some o_g belongs to *GO* (noted $o_g \in GO$), a territory *Terr* can be either formalized by *Terr*∈ *Earth* or by *Contains* (*Earth, Terr*), both having different semantics, but practically both equivalent; so we can state:

$$(Terr \in Earth) \equiv (Contains\ (Earth,\ Terr)) \equiv (Inside\ (Terr,\ Earth))$$

5.2.2. *Projective and other spatial relations*

But other relations exist such as projective (or cardinal such as *North/South, East/West*) relations and distance *(near/far)* relations (Figure 5.14).

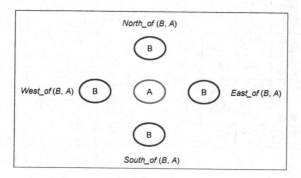

Figure 5.14. *Projective and distance relations*

This definition is valid for points and small area. However for huge areas, it must be redefined. See section 5.3.3.

5.2.3. *Rectangle relations*

Lee and Hsu [LEE 90, LEE 92] proposed a table (Figure 5.15) representing all spatial relations between two rectangles. They found a total of 169 types in which they number: 48 disjoint, 40 join, 50 partial overlaps, 16 contains and 16 belongs (corresponding to inside). Each of them has a symbol: for instance, the three first relations in the upper row are noted respectively, <<, /<* and /*<*.

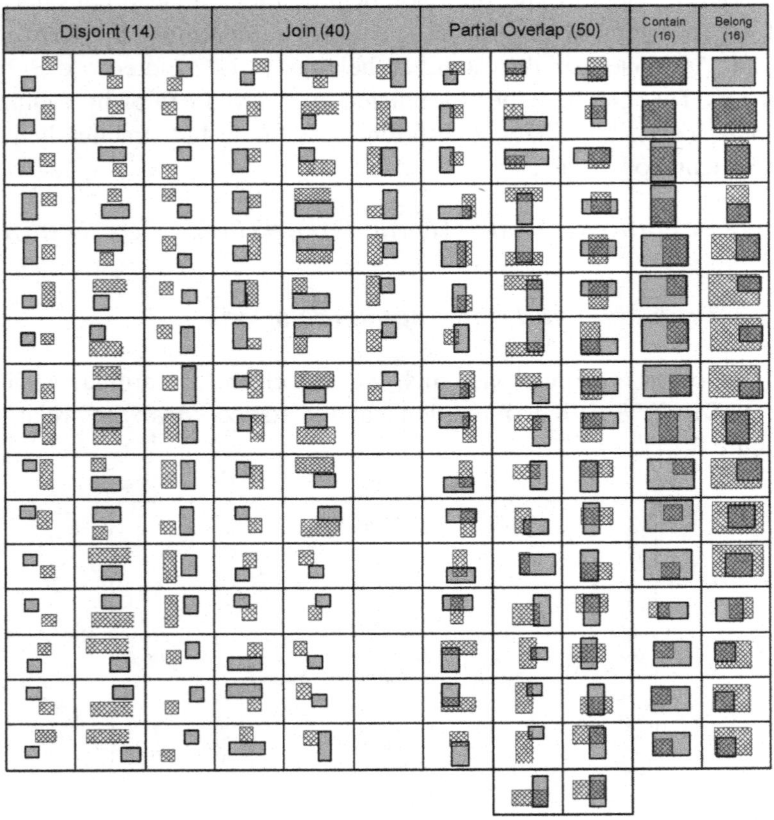

Figure 5.15. *Lee and Hsu's relations [LEE 90, LEE 92]*

5.3. Characteristics of spherical spatial relations

Let us remember that each territory *Terr* is inside the *Earth*, so we can write *Inside (Terr, Earth)* and *Contains (Earth, Terr)*.

5.3.1. *Projective relations*

As an example, let us consider the relation "*North*" and its transitivity, the following rule holds.

$\forall A, B, C \in Earth,$ G-*Type* $(A) = Point$, G-*Type* $(B) = Point$, G-*Type* $(C) = Point$: *North* $(A, B) \wedge North (B, C)$ $\Rightarrow$ *North* (A, C)	Rule 5.1

For South, a similar rule can be written, but for East and West, this is different due to the Earth's curvature. Let us first consider the following:

$\forall A, B, C \in Earth,$ G-*Type* $(A) = Point$, G-*Type* $(B) = Point$, G-*Type* $(C) = Point$: *East* $(A, B) \wedge East (B, C) \wedge (Longitude (C) - Longitude (A) < 180)$ $\Rightarrow$ *East* (A, C)	Rule 5.2

and its complementary rule:

$\forall A, B, C \in Earth,$ G-*Type*$(A) = Point$, G-*Type* $(B) = Point$, G-*Type* $(C) = Point$: *East* $(A, B) \wedge East (B, C) \wedge (Longitude (C) - Longitude (A) > 180)$ $\Rightarrow$ *West* (A, C)	Rule 5.3

5.3.2. *Projective relations for areas*

As projective relations are easy to define for points, they are more complex for areas. Do not forget that areas can be disconnected such as countries with several islands. Moreover, some countries can have holes such as Italy with Vatican City and San Marino. Some are constituted of pieces of territories which are very far from the main component: for instance France with Martinique, Guyana, New Caledonia, etc.

Let's consider Canada, conterminous states of the United States of America (USA for short) and Mexico. As it is easy to claim the Mexico is south of Canada, what is the exact projective relation between USA-Mexico and Canada-Mexico? It is common to claim that "Canada is north of the conterminous states of the USA", but Canadian cities such as Ottawa or Toronto are south of Seattle. A solution is to consider centroids of those areas. So but taking this definition, we can claim:

North ("Canada", "USA")

North ("USA", "Mexico").

Moreover, one can claim the *East* ("Mexico", "USA"). On the contrary, is Canada east or west of the USA? By using both centroids, there is an answer, but this answer is not totally convincing.

To determine the relative projective position of two areas, namely A and B, for instance to state *North*(A, B), *South*(A, B), *West*(A, B) or *East*(A,B), the best solution is based on the comparison between MBR. However, it is also a matter of definition firstly in natural language. As it is simple to state that Paraguay is south of Venezuela, it is not clear when compared to Brazil: some people can say south or others west; the great majority can affirm south-west. But the big problem concerns non-connected places. If you want to state a projective relations between USA as a whole (including all oversea territories spread across Caribbean and the Pacific) the Commonwealth of Nations with its 53 member states (including among others Australia, Canada, etc.), it is impossible to claim which is west and which is north.

The second aspect is inside or outside. Indeed it is common to state that Quebec is east of Canada, or the Côte d'Azur is south of France.

For connected territories, two cases must be distinguished based on the spatial relationships between *A* and *B*, namely *Touches* (*A*, *B*) or *Disjoint* (*A*, *B*). Let us denote respectively *MBR* (*A*), *MBR* (*B*) the minimum bounding rectangles and *Centroid* (*A*) and *Centroid* (*B*) their respective centroids (Figure 5.16).

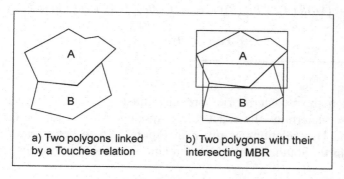

a) Two polygons linked by a Touches relation

b) Two polygons with their intersecting MBR

Figure 5.16. *Determining a* North *relation between two polygons by means of their Minimum Bounding Rectangles*

In *Touches* case, there is an intersection between *MBR* (*A*) and *MBR* (*B*) as illustrated in Figure 5.17.

∀*A*, *B* ∈ *Earth*, G-*Type* (*A*) = *Area*, G-*Type* (*B*) = *Area*: *Touches* (*MBR* (*A*), *MBR* (*B*)) ∧ *North* (*Centroid* (*A*), *Centroid* (*B*)) ⇒ *North* (*A*, *B*)	Rule 5.4
∀*A*, *B* ∈ *Earth*, G-*Type* (*A*) = *Area*, G-*Type* (*B*) = *Area*: *Touches* (*MBR* (*A*), *MBR* (*B*)) ∧ *South* (*Centroid* (*A*), *Centroid* (*B*)) ⇒ *South* (*A*, *B*)	Rule 5.5

$\forall A, B \in Earth,$ G-*Type* (A)=*Area*, G-*Type* (B)=*Area:* *Touches* (MBR (A), MBR (B)) $\wedge$ *West* (*Centroid* (A), *Centroid* (B)) $\Rightarrow$ *West* (A, B)	Rule 5.6
$\forall A, B \in Earth,$ G-*Type* (A)=*Area*, G-*Type* (B)=*Area:* *Touches* (MBR (A), MBR (B)) $\wedge$ *East* (*Centroid* (A), *Centroid* (B)) $\Rightarrow$ *East* (A, B)	Rule 5.7

When the two territories are disjointed, we let the reader as an exercise, determine the exact rules. Another solution can be based on Lee and Hsu relations as given in Figure 5.15 to determine rules to state North, South, East and West relations between territories.

5.4. Spatial relations in urban space

Considering a city and spatial relations between urban objects, we can make the assumption that the conventional 3D Cartesian is valid. Moreover, considering again streets, some observations can be made:

– they are one-way streets and sometimes two-way streets can have several lanes;

– some objects are positioned on the street (zebra crossings), some underneath such as sewerages, and some above such as traffic lights;

– some concrete concepts such as sidewalks, medians, crossroads, *T*-junction, roundabouts, road signs, curves and engineering networks can be defined with the "*has_a*" semantic relation, but their "topological semantics" are stronger;

– some objects such as engineering networks can be located under streets or under sidewalks;

– as previously stated, for some actors, streets are defined by the lines with parcels whereas for others the streets are reduced to the asphalted part (Figure 3.1).

So those observations imply that Allen or Egenhofer relations are not sufficient to describe relationships between street objects. So, the question is "what could be the minimum set of useful geographic relations?"

5.4.1. *Other binary topological relations*

on (*street, pedonal_zebra*)

underneath (*street, sewerage*)

above (*street, traffic_light*)

along (*sidewalk, street*)

on (*sewerage_grid, street*).

5.4.2. *Relations between urban features and places*

host (*barrack, army*)

host (*hospital, health_activity*).

5.5. Ribbon operations and relations

As previously enumerated, ribbons are derived from longish rectangles. So the relations between areas can be applied to ribbons. However due to their particular shapes, other interesting relations between ribbons can be detailed. First let us examine basic operations and some new relations.

5.5.1. *Simple operations and relations*

Two operations can be defined. Considering that any ribbon can be decomposed into sub-ribbons, either longitudinally or laterally, we can define two operations, longitudinal splitting and lateral splitting (Figure 5.17). Of course, those operators can be recursively used.

For longitudinal splitting, one has $w_1=w_2=w$ and $l_1+l_2=l$, whereas for lateral splitting $w_3+w_4=w$ and $l_3=l_4=l$.

A solution can be to use Lee and Hsu relations as given in section 5.2.3. More, due to the semantics of ribbon, a lot of previous relations can be discarded. This is not sufficient. Suppose a road which alternates between simple and dual carriageways. In this case, we need to consider three ribbons, corresponding to dividing and merging. Finally, Figure 5.18 gives the more interesting ribbon relation, namely *Side-by-Side*, *Edge-to-edge*, and merging respectively rectangular ribbons, ribbons and loose ribbons. Similarly, other relations can be defined, for instance crossings, *T*-junctions, etc. Examples along a street and a river bank are given respectively in Figures 5.19a and b.

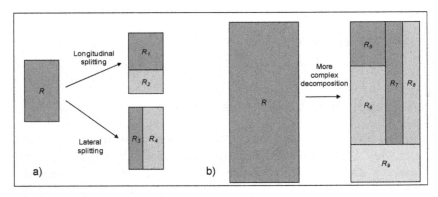

Figure 5.17. *Two ribbon operators, longitudinal splitting and lateral splitting. a) Definitions; b) A more complex decomposition*

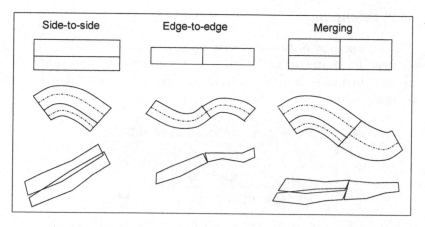

Figure 5.18. *The more interesting relations between ribbons*

For instance, in transportation and along rivers, the following relations can hold:

SIDE_BY_SIDE (*Platform, railways*)

SIDE_BY_SIDE (*Bus_stop, Bus_lane*)

SIDE_BY_SIDE (*Levee, River*)

SIDE_BY_SIDE (*Towpath, River*).

Initially towpaths along rivers were made for horse-drawn boats; but more and more they are replaced by bike lanes. Examples are given Figure 5.19.

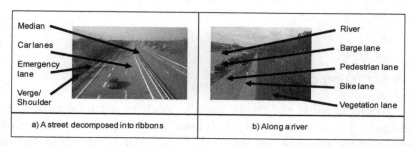

Figure 5.19. *Decomposition into ribbons. a) A street; b) Along a river*

5.5.2. *Orientation*

Since one-way or two-way streets exist, orientation can be defined whereas for some cases of ribbons orientation is not valid. For instance, in a conventional street, the decomposition into ribbons is as in Figure 5.20.

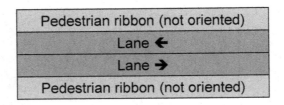

Figure 5.20. *Ribbon orientation*

5.5.3. *3D Relations between ribbons*

In the case of engineering networks, it could be interesting to define 3D relations between ribbons. Indeed consider sewerage, gas and water pipes. In most cases, they are buried underground. These under- and aboveground relations can be defined as illustrated Figure 21.

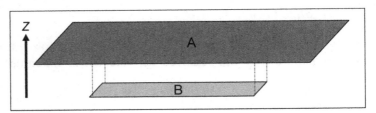

Figure 5.21. *Example of 3D relations for ribbons*

5.5.4. *Chaining ribbons*

To model roads and rivers, usually polylines are used to describe the axis; but sometimes two polylines can be used to model river banks or the extremity of the road. As a consequence those feature representations can be transformed into ribbons with different widths. For several other reasons, one can have a set of different ribbons that must be concatenated to form a chain of ribbons. Figure 5.22 gives an example (Figure 5.22(a)) of several ribbons transformed into a chain of ribbons (Figure 5.22(b)). In Figure 5.22(c), a case is presented needing an additional area to join the sides of two ribbons. This can be done by using morphology mathematics, namely, dilation and erosion operators [SER 82].

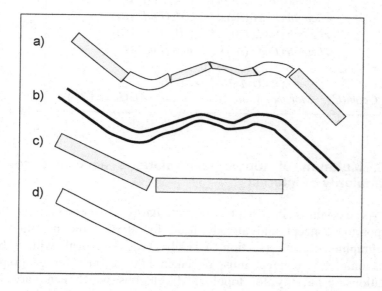

Figure 5.22. *Chaining ribbons: a) An example of different ribbons;*
b) Chain of ribbons; c) A case of two rectangular ribbons;
d) Additional curves to join the sides

Consider two regular ribbons. We can chain them by applying the following rule

$\forall A, B \in GO, \exists C \in GO,$ G-*Type (A)* = *Ribbon,* G-*Type (B)* = *Ribbon,* G-*Type (C)* = *Ribbon,* *END_TO_END (A, B):* *Width (A)* = *Width (B)* $\Rightarrow$ *Geom(C)* = *Union (Geom (A), Geom (B))*	Rule 5.8

Concerning loose ribbons, the previous rule is generalized into the following:

$\forall A, B \in GO, \exists C \in GO,$ G-*Type (A)* = *Ribbon*, G-*Type (B)* = *Ribbon*, G-*Type* *(C)* = *Ribbon*: *Length(End1 (A))* ⋈ *Length (End1 (B))* ∨*Length (End1 (A))* ⋈ *Length (End2 (B))* ∨*Length (End2 (A))* ⋈ *Length (End1 (B))* ∨*Length (End2 (A))* ⋈ *Length (End2 (B))* $\Rightarrow$ *{END_TO_END (A, B)*; *Geom(C)* = *Morpho_Math_Union (Geom (A), Geom (B))}*	Rule 5.9

5.6. Mutation of topological relations according to the granularity of interest

As explained in Chapter 3, granularity of interest is a very important concept and already used for studied the mutations of geographic objects (section 4.4). In order to deal with robust reasoning, this concept must be taken into account for geographic relations, especially for topological relations which can vary. For instance, it is common to claim that a road is going along the sea, so implying a *Touches* relation between the road and the sea. But if we consider carefully, sometimes there are small beaches between the road and the sea (Figure 5.23). From a cartographic point of view, the type of relation will vary: indeed at a scale of 1:1,000, the relation is *Disjoint* whereas at 1:100,000, there is a *Touches*. More generally, the concept of granularity of interest will enlarge the concept of scale: it is clear that the interest for the same zone for a local politician and a national politician can be different.

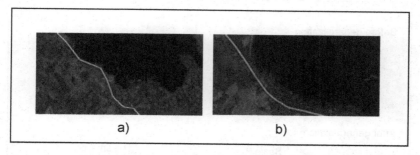

Figure 5.23. *According to scale, the road* Touches *or not the sea*

5.6.1. *Example of topological mutation due to granularity of interest*

As a consequence at one scale, the road *Touches* the sea, but at another scale at some places, this is a *Disjoint* relation, the general mechanism of which is illustrated Figure 5.24. Let us consider two geographic objects O^1 and O^2 and O_σ^1 and O_σ^2 their cartographic representations, for instance the following assertion holds:

$\forall O^1, O^2 \in GO, \forall \sigma \in Scale,$ $O_\sigma^1 \equiv 2\,Dmap(O^1), O_\sigma^2 \equiv 2Dmap(O^2), Disjoint\,(O^1, O^2):$ $Dist\,(O^1, O^2) < \varepsilon^2$ $\Rightarrow$ $Touches\,(O_\sigma^1, O_\sigma^2).$	Rule 5.10

Similar assertions could be written when *Contains, Overlap* relationships. In addition, two objects in the real world with a *Touches* relation can coalesce into a single one.

As a consequence, in considering what is true at one scale, can be wrong at another scale. So, any automatic system must be robust enough to deal with this issue.

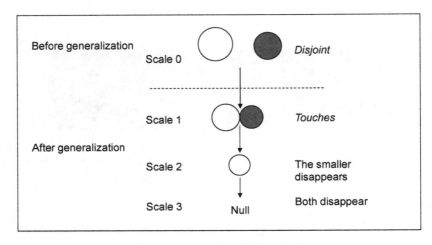

Figure 5.24. *The mutation Disjoint-to-Touches*

From-to mutation	Initial scale	Smaller scale
Overlap ∧ *C1* ⇒ *Equal*		
Disjoint ∧ *C2* ⇒ Touches	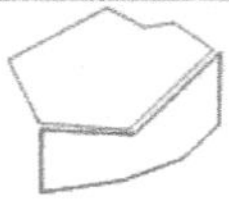	
Overlap ∧ *C3* ⇒ *Touches*	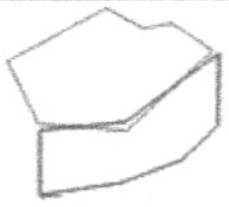	
Overlap ∧ *C4* ⇒ *Covers*	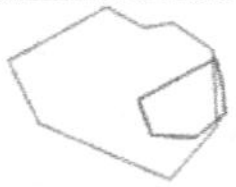	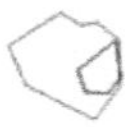
Contains ∧ *C5* ⇒ *Covers*	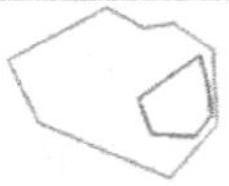	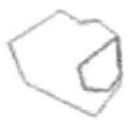

Figure 5.25. *Some mutation rules for topological relations*

5.6.2. *Mutation table of Egenhofer relations*

As consequence, a bunch of mutation rules for topological relations can be identified not only according to scale, but also according to the shapes of the concerned geo-objects. Examples of other mutations rules are given in Figure 5.24 in which C_i denote some possible additional conditions.

5.6.3. *Mutation of ribbon relations*

Similarly, some mutation relations can be created for ribbons. For instance, consider a narrow ribbon located between two ribbons, it can be discarded according to some thresholds ε_s and ε_e. See examples in Figure 5.26.

$\forall A, B, C \in Earth,$ G-*Type* $(A) = Ribbon$, G-*Type* $(B) = Ribbon$, G-*Type* $(B) = Ribbon$: $SIDE\_BY\_SIDE\,(A, B) \wedge SIDE\_BY\_SIDE\,(B, C)$ $\wedge\ width\,(B) < \varepsilon_s$ $\Rightarrow$ $SIDE\_BY\_SIDE\,(A, C)$	Rule 5.11
$\forall A, B, C \in Earth,$ G-*Type* $(A) = Ribbon$, G-*Type* $(B) = Ribbon$, G-*Type* $(B) = Ribbon$: $EDGE\_TO\_EDGE\,(E, F) \wedge EDGE\_TO\_EDGE\,(F, G)$ $\wedge\ length\,(F) < \varepsilon_e$ $\Rightarrow$ $EDGE\_TO\_EDGE\,(F, G)$	Rule 5.12

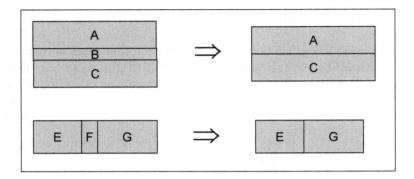

Figure 5.26. *Discarding narrow ribbons. a) case SIDE_BY_SIDE; b) case EDGE_TO_EDGE*

5.7. Other geographic relations

In addition to the previously mentioned relations, there are many other types of geographic relations between geographic objects. Let us mention a few:

– about rivers: tributary, downstream, upstream, estuary;

– about roads: crossing, *T*-junctions, triangular junction, parallel;

– about cities: near, twin cities;

– about cities and country: belong, capital;

– within cities, social services, school, transportation, energy, metro-lines, etc.;

– about any territory: neighbor;

– about vegetation, biotopes;

– in civil engineering concerning bridges, tunnel, engineering networks;

– etc.

All those relations not only can be defined, but also used in geographic reasoning.

5.8. Conclusion regarding geographic relations

The great majority of papers in GIS claims to deal with spatial reasoning and few on geographic reasoning (see their keywords for instance). In this chapter, starting from conventional spatial relations, several geographic relations were presented as a basis for geographic reasoning.

Emphasis was given on mutation of geographic relations: we think this is important to allow some flexibility of geographic reasoning regarding scale; in other words, this mechanism allows independence from scale which is a salient objective of geographic reasoning.

Finally, relations in geographic knowledge bases must be modeled as follows.

– binary relations such as projective relations (North, etc.) being so easy to determine and very numerous, let me propose not to store them, but to generate them on-demand;

– other binary or n-ary relations can be stored directly.

More precisely, relationships will be stored according to the following model:

$Relation\ (GeoID_1,\ GeoID_2,\ ....,\ GeoID_n).$

However, there are more complex relations linking several geographic objects, which will be analyzed in Chapter 7. However, before this, it is necessary to study ontologies because they integrate new concepts which are of importance to constitute complex geographic objects.

6

Geographic Ontologies

This chapter presents the concept of ontology and more exactly the concepts of geographic ontologies. Those ontologies model not only conventional geographic features with their semantic relations, but also with geographic relations which exist between those features. The scope of this chapter will be to identify those spatial relations, to show how to use them for modeling and manipulating geographic ontologies.

As a preliminary example, Figure 6.1 gives a small ontology for natural disasters with only two relations, "*is_a*" and "*implies*" modeling causalities between some natural disasters.

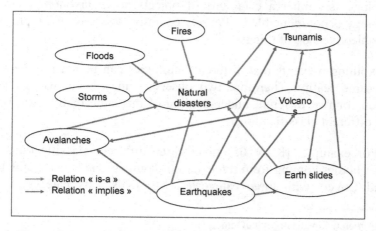

Figure 6.1. *An example of a small ontology for natural disasters*

6.1. Introduction

The word "ontology" comes from the ancient Greek "οντος" (being) and "λογια" (discourse), that is the discourse about existing objects. This word, usually written with a capital "O" is mostly used in philosophy and theology (God = "I Am that I Am"). In information technology, an ontology (with a lower case "o") refers to modeling things existing in the discourse which is a fundamental idea in data modeling: when something has no name, it is not existing in our mind, so not existing in our culture, so not existing in our world. It is said that Eskimos have 19 words to describe "snow" whereas some Equatorial tribes may have no word for this natural phenomenon; so the concerned ontologies, if any, include different concepts perhaps due to climate conditions.

Aristotle[1] defines ontology as "the theory of things and their relations", or as "the theory of entities, especially those existing in language". In information technology [GRU 93], the more used definition is "an ontology is a specification of a conceptualization", so on ontology is an artifact created to describe the meaning of vocabulary. Indeed Guarino [GUA 98] said that, in artificial intelligence, an ontology represents an artifact made with a vocabulary for building reality, accompanied with a set of implicit assumptions concerning the meaning of words and of the vocabulary. So an ontology is neither a catalogue of objects nor a taxonomy, but an ontology is not reducible to a purely cognitive analysis, and represents the objective side of things.

Nothing prevents that different ontologies can be used to describe the same reality. Therefore two observers may have two different visions or two different understanding of the same reality, so giving two different classifications.

For example, [KAV 05] gives three different classifications for water bodies coming from three various sources (Table 6.1), CORINE Land Cover[2] with three categories, MEGRIN[3], six categories and

---

1 https://en.wikipedia.org/wiki/Ontology.
2 http://www.eea.europa.eu/.

WordNet[4], seven categories. In other words, facing the same water body, it can be categorized differently according to those systems.

Finally, an ontology can be considered as a conceptualization method; the main idea is to replace the domain of semantic interpretation (= conceptualization) by an ontology. Then, an immense description in intension must be built with few rules, integrating all possible and plausible facts organized in domains, contexts and applications. But the big question is "where can we find all these concepts?". For example, there is no authority able to define what a "seat" is!

Ontologies	Category type
CORINE Land Cover	Peat bog Water course Water body
MEGRIN	Bog Canal Lake/pond Salt marsh Salt pan Watercourse
WordNet	Body of water Bog Canal Lake Pond Salt pan Watercourse

Table 6.1. *Example issued from [KAV 05]*

6.2. Generalities about ontologies

Ontologies[5] are, as a solution to the problem discovered by users of Internet, the biggest problem after the emergence of the phenomenon

---

3 http://www.megrin.org/gddd/gddd.htm.

4 https://wordnet.princeton.edu/.

5 A preliminary version was proposed in [LAU 16a].

of silence and noise from information retrieval. Silence means that existing information cannot be accessible whereas the noise phenomenon is the situation when we obtain undesirable or unwanted information. The main cause of the approach is based on syntax and on semantics.

6.2.1. *Role and definition*

Ontologies were developed in artificial intelligence to facilitate knowledge sharing and reuse. Since the beginning of the 1990s, ontologies have become a popular research topic investigated by several artificial intelligence research communities, including knowledge engineering, natural-language processing and knowledge representation. In the database community, ontologies are also widely used for modeling and for interoperability. More recently, the notion of ontology is also becoming widespread in fields such as intelligent information integration, cooperative information systems, information retrieval, electronic commerce, and knowledge management. The reason for which ontologies are becoming so popular is in large part due to what they promise: a shared and common understanding of some domains that can be communicated between people and application systems. Ontologies are developed to provide machine-processable semantics of information sources that can be communicated between different agents (software and humans).

Many definitions of ontologies have been given in the last decade, but one that, in our opinion, best characterizes the essence of an ontology is based on the related definitions in [GRU 93]: remember that an ontology is a formal, explicit specification of a shared conceptualization. A "conceptualization" refers to an abstract model of some phenomenon in the world which identifies the relevant concepts of that phenomenon. "Explicit" means that the type of concepts used and the constraints in use are explicitly defined. "Formal" refers to the fact that the ontology should be machine-readable. "Shared" means that several actors must reach a consensus.

6.2.2. *Categories of ontologies*

Depending on their generality level, different types of ontologies may be identified that fulfill different roles in the process of building a knowledge base system [GUA 98, VAN 97]. Among others, we can distinguish the following ontology types:

– *domain ontologies* capture the knowledge valid for a particular type of domain (e.g. electronic, medical, mechanic, digital domain);

– *metadata ontologies* like Dublin Core [WEI 95] provide a vocabulary for describing the content of online information sources;

– *generic or common sense ontologies* aim at capturing general knowledge about the world, providing basic notions and concepts for things like time, space, state, event etc. [FRI 97]. As a consequence, they are valid across several domains. For example, an ontology about mereology (*part_of* relations) is applicable in many technical domains [BOR 97];

– other types of ontology are so-called *method and task ontologies* [FEN 97, STU 96]. Task ontologies provide terms specific for particular tasks (e.g. "hypothesis" belongs to the diagnosis task ontology), and method ontologies provide terms specific to particular Propose-and-Revise method ontology). Task and method ontologies provide a reasoning point of view on domain knowledge.

6.2.3. *Ontology approaches*

There are five approaches to ontology design: inspirational, deductive, synthetic, collaborative and inductive [HOL 02]:

a) *The inspirational approach:*

In the inspirational approach, a designer takes decisions alone to gather the terms of the domain analysis, design and verification of ontology. The developer must be both a domain expert and an expert in ontology design to ensure the success of the design, and above all, to ensure the adoption of the ontology by the user community. This process depends heavily on the creativity of one person, his/her inspiration and his/her personal perception of the area.

b) *The deductive approach:*

With a deductive approach, the general principles are first adopted and then processed and applied to the target domain. The resulting ontology can be seen as an instance object of these general concepts.

c) *The synthetic approach:*

In the synthetic approach, a set of related ontologies is identified. The developer then synthesizes the elements of these ontologies with the concepts of the new target area, producing a new unified ontology.

d) *The collaborative approach:*

The mark of a "modern" ontology is its large size and high complexity. This kind of ontology encompasses a rich set of knowledge that its understanding exceeds that of any single developer or designer or even a small team of designers. The development of a large-scale ontology must be the fruit of a joint effort of several domain experts and software designers.

e) *The inductive approach:*

With the inductive approach, ontology is developed by observing, examining, and analyzing a specific case or cases in the domain of interest. The characterization of the resulting ontology for a specific case is applied to other cases in the same field. The design is based largely on the widespread cases selected during development.

6.2.4. Ontology examples

Several projects of constructing ontologies exist. Among them, let us mention WordNet[6], CYC[7] and TOVE[8]. For instance, WordNet [FEL 99] is an online lexical reference system whose design is

6 https://wordnet.princeton.edu/.
7 http://www.cyc.com/.
8 http://www.eil.utoronto.ca/theory/enterprise-modelling/tove/.

inspired by current psycholinguistic theories of human lexical memory. English nouns, verbs, adjectives and adverbs are organized into synonym sets, each representing one underlying lexical concept. Different relations link the synonym sets. It was developed by the Cognitive Science Laboratory at Princeton University. WordNet now contains around 100,000 word meanings organized within a taxonomy.

WordNet groups words into five categories: noun, verb, adjective, adverb and function word. Within each category it organizes the words by concepts (i.e. word meanings) and via semantic relationship between words. Examples of these relationships are:

– synonymy: similarity in meaning of words, which is used to build concepts represented by a set of words;

– antonymy: dichotomy in meaning of words, mainly used for organizing adjectives and adverbs;

– hyponymy: *is_a* relationship between concepts. This *is-a* hierarchy ensures the inheritance of properties from superconcepts to subconcepts;

– meronymy: *part_whole* relationship between concepts;

– morphological relations which are used to reduce word forms.

Note that there is a topological similarity for the *Part_whole* relation applied to territories. Indeed it can correspond both to *Contains* and *Covers*.

Part_whole (*Earth, Terr*) $\Rightarrow$ *Contains* (*Earth, Terr*) $\lor$ *Covers* (*Earth, Terr*)

Take the example of airports as all they have a take-off runway. From a set-theoretical approach, we must consider a *Part_whole* relation whereas from a topological point of view, we have either a *Contains* relation or a *Covers* one.

6.2.5. Ontology components

An ontology consists of a number of different components. The names of these components are dependant on the expressivity of an ontology (or, in general, of a knowledge representation language) used. Despite this, their core components are largely shared between different ontologies. The main components of ontology are: concepts, instances, and relations [LOR 10].

a) Concepts:

Concepts, also called Classes or Types, are a core component of most ontologies. A Concept represents a group of different individuals that share common characteristics, which may be more or less specific. For instance, *Person* is a Concept that represents a set of individuals (persons). One concept may be a sub-concept (also known as subclass, or kind) of another concept; this means that if the Concept *C'* is a sub-concept of *C*, then any individual of *C'* will also be an individual of *C*. Concepts may also share relationships with each other; these relationships describe the way individuals of one Concept relate to the individuals of another.

b) Instances:

Individuals also known as instances or particulars are the base unit of an ontology. They are the things that the ontology describes or potentially could describe. Individuals may model concrete objects such as people or machines; they may also model more abstract objects such as countries, person's job or a function.

c) Relations:

Relations in an ontology describe the way in which individuals relate to each other. Relations can normally be expressed directly between individuals, for example, the relation *hasSibling* (Figure 6.2) might link the individual Matthew to the individual Gemma, or between Concepts, for example; the relation *livesInCountry* might link the concept Person to the concept Country. In the latter case, this describes a relationship between all individuals of the Concepts.

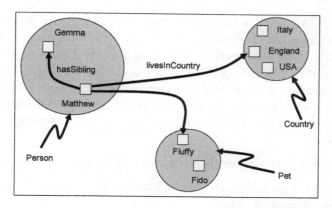

Figure 6.2. *Example of an ontology including instances*

6.2.6. *Ontology languages*

The literature offers many description languages to express ontologies, based on different representations. According to [GOM 02], some of them are based on XML syntax, such as Ontology Exchange Language (XOL), Resource Description Framework (RDF) and RDF Schema, and OWL (Web Ontology Language) are languages created by World Wide Web Consortium (W3C)[9] working groups (Figure 6.3) [DJU 06].

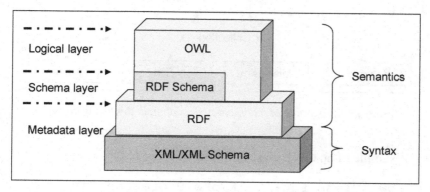

Figure 6.3. *Ontology languages [DJU 06]*

---

9 https://www.w3.org/.

OWL [SMI 04] is a key to the semantic web that was proposed by the Web Ontology Working Group of W3C. It is a language extension of RDF Schema (see section 2.3). It is a general purpose ontology language that contains all the necessary constructors to formally describe most of the information management definitions: classes and properties, with hierarchies, and also range and domain restrictions. Basic Terminology OWL has the power of expressing richer properties:

– symmetric properties (If A connects B then B also connects A);

– transitive properties (If A is included in B and B is included in C then A is contained in C);

– functional properties (At most one value for the property);

– inverse properties (if A is related to B in a way of relation X, then B is related to A in a way of relation Y; so X and Y relations are inversed.

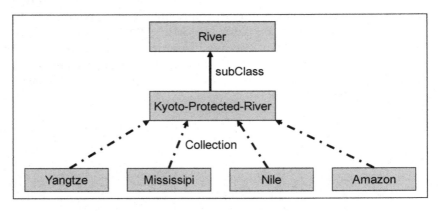

Figure 6.4. *Example of an ontology described in the text with OWL*

Example of an OWL class definition (Figure 6.4):

```
<owl:Class rdf:ID="Kyoto-Protected-River">
<rdfs:subClassOf rdf:resource="#River"/>
<owl:oneOf rdf:parseType="Collection">
```

```
  <geo:River
rdf:about="http://www.china.org/rivers#Yangtze"/>
  <geo:River
rdf:about="http://www.us.org/rivers#Mississippi
"/>
  <geo:River
rdf:about="http://www.africa.org/rivers#Nile"/>
  <geo:River  rdf:about="http://www.s-america.org/
rivers#Amazon"/>
  </owl:oneOf>
  </owl:Class>
```

Restrictions in property definitions:

– onProperty: specifies that on which property the restriction will be applied;

– allValuesFrom: specifies which values are accepted by the property;

– hasValue: specifies which value is the property has to be exactly;

– someValue: specifies that the property has to have at least a value;

– cardinality: specifies the occurrence of the property;

– minCardinality: specifies the minimum occurrence of the property;

– maxCardinality: specifies the maximum occurrence of the property;

To conclude this section, one can say that creating an ontology is an approach similar to conceptual data modeling. But in addition, it can include constraints, business rules, derived rules, etc. There is no consideration about storage, but a distinction between concepts and terms must be made. Finally, from a mathematical point of view, an ontology is a graph of concepts, so a semantic network.

6.2.7. *Conventional ontologies applied to geography*

Geographic ontologies can be classified into three categories [CUI 03]:

– ontologies of space are specifically dedicated to the description of concepts that characterize space as point, line, etc. These ontologies are typically developed by leading bodies of standardization;

– ontologies of geographic area can model the concepts of hydraulic data, an ontology with concepts of urban data, or an ontology describing the concepts of electrical data networks, etc. These are "domain" ontologies developed by the user community in their concerning field;

– the spatial ontologies (or spatio-temporal), which are ontologies whose concepts are located in space. A temporal component is often needed as an addition to the modeling of geographic information, because geographic applications also often wield temporal or spatiotemporal data.

Geographic ontologies have specific needs [CUI 03] that are related to the necessity to define spatiality by using various types of spatial data (line, point, or surface, etc.), types of space objects (i.e. objects with spatial attributes), spatial relations as topological relationships and/or continuous fields (raster) intentionally define spatial concepts by using axioms containing spatial predicates and reason on the spatiality of instances objects i.e. inferred from the spatial relationships describing the set of valid spatial relationships.

For illustrating the difficulties to classify geographic objects, let us note that Table 6.2, issued from [KAV 05], depicts how in some existing systems, water bodies are described. Table 6.3 gives an excerpt of littoral geographic objects cited by [RAP 02].

Obj. No.	Real World phenomena - Source Terminology	Object name	Object Description			
			Attributes		Object Identity	Implementation
			Source Terminology	User Defined Attributes		Point, Line, Area or Samples
34	Coastline / Shore / Shoreline / Shoreline movement and configuration / Mean low water / Mean low water (springs) / Median low water mark / Low water mark / Low water (mean) / Low water (spring) / Lower tidal limit	OBJECT-COAST (MLWs - MHWs)	Heritage Coast / Coastline (managed) / Coastline (unspoilt) / Natural Coastline / Coast (undeveloped) / Coast (restored) / Coastline (rural) / Urban coasts	Heritage Coast / Developed / Undeveloped	PHYSICAL ENTITY : SPATIALLY HOMOGENEOUS	VECTOR (AREA) MULTI-ATTRIBUTE
35	Point Of Closure / Base line	OBJECT-CLOSURE		Depth	COGNISED ENTITY : SPATIALLY HOMOGENEOUS	VECTOR (LINE) : SINGLE ATTRIBUTE
36	Areas of Responsibility / Administrative Boundaries / Admin. / County Boundaries / Coastal Jurisdiction	OBJECT-ADMINISTRATION	Coastal cells (management) / CZM unit / Sea surface management areas / Buffer zones / Land ownership	Management Zone / Land Ownership	GEOPOLITICAL ENTITY : SPATIALLY HOMOGENEOUS	RASTER : MULTI-ATTRIBUTE

Table 6.2. *Excerpt of a geographic ontology for coastline objects [RAP 02]*

	Street cleaners	Postmen	Electricity company
Private roads	No	Yes	??
Public streets	Yes	Yes	Generally yes
Streets with electricity	?	?	Yes
Streets without electricity	?	?	No
Total	234	251	241

Table 6.3. *For the same city, different organizations have a different number for streets. As a consequence, definitions should be different as they call to be complete*

As an additional example, let us consider streets. Every three-year-old child knows what a street is, especially when their mother says "be careful when crossing the street!". So, if one wants to set out a list of streets, one could ask a street-cleaner, a postman and an electrician.

All of them will claim "Yes, we have the street on file!". But by examining those lists (Table 6.3), one can observe that they are not identical: the street-cleaner only cleans public streets, the postman only works in residential streets and the electrician only works on streets with electricity. This example shows that there are various categories in urban features: firstly, categories bearing the same names (streets) belong to the same category, but secondly, they are different because different actors have different interests.

6.3. Towntology: ontologies for urban planning

Under the name of Towntology, there were two projects. The first one [KEI 06a] was initiated in Lyon in 2002 in order to create an ontology for urban planning. Since this task was immense, a European project [TEL 07] was launched in order to involve other European laboratories (COST C21[10]). The main objective of this Action was to increase the knowledge and promote the use of ontologies in the domain of Urban Civil Engineering projects, in the view of facilitating the communications between information systems, stakeholders and UCE specialists at the European level.

6.3.1. *Genesis and objectives of the towntology project*

The first Towntology project was initiated between two laboratories at INSA Lyon, one in computing (LIRIS) and one in urban planning (EDU). The EDU laboratory was in charge of developing and populating the ontology, whereas the computing laboratory was in charge of defining the structure and designing all software modules [KEI 06a]. Initially, the objectives were as follows:

– to define a pre-consensus ontology, that is storing various textual definitions, possibly multimedia definitions;

– to access visually to the system;

---

10 http://www.cost.eu/COST_Actions/tud/C21.

– to build several semantic networks;

– to integrate photos, drawings and sound (for instance for traffic noise);

– to integrate metadata for the origin and traceability of definitions;

– and to make updating possible.

But in the first months, it was clear that for software developers, the role of ontologies was to help interoperability between urban databases whereas for town planners, clarification of the vocabulary was seen as the main objective.

Initially, the goal was to create an ontology for street management (Figure 6.5). So more than 800 concepts were integrated along with nine relations: (*i*) is made of, (*ii*) is composed of, (*iii*) is located in, (*iv*) is used for, (*v*) is located on, (*vi*) is a, (*vii*) is a subset of, (*viii*) depends on and (*ix*) is a tool for. A second ontology was made for analyzing mobility.

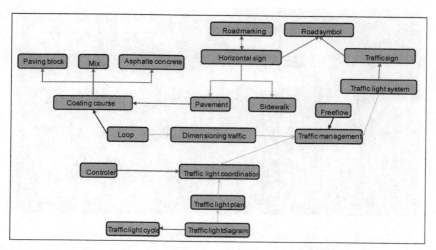

Figure 6.5. *Excerpt of a Towntology ontology for street management*

The Towntology software allows for the construction and the visualization of a semantic network of concepts. Thus several natural language definitions or pictorial illustrations can be associated to a unique concept in order to express all possible interpretations. Ontologies constructed by this tool are classified as lightweight ontology because no reasoning process is actually associated. Thus their expressiveness is limited.

The Towntology browser offers three types of access to a concept [KEI 06a] (Figure 6.6):

1) select a term, which is a concept label in an alphabetical ordered list;

2) navigate in a graph representing the semantic network in order to find the appropriate concept thanks to its relations with other concepts;

3) select part of images annotated by concepts;

4) all the ontologies are stored into an XML file specifically designed.

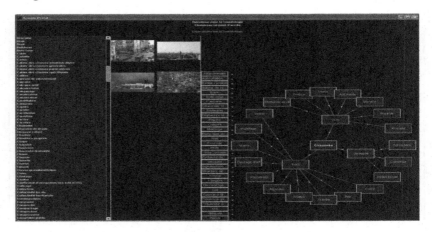

Figure 6.6. *Initial Towntology portal; from left to right, textual access, visual access, graph-based access*

6.3.2. *Lessons learnt from the towntology project*

After the Lyon experience, a European project involving a dozen European laboratories was launched with the following objectives:

– cover the whole urban field, each part assigned to a laboratory;

– find a consensus for each definition;

– create some tool to reach the consensus;

– develop in parallel several sub-ontologies referring each other;

– check consistency;

– consolidate the various sub-ontologies;

– check completeness;

– take multiplicity of languages into account;

– take legislative context into account;

– study encoding languages such as OWL, Descriptive Logics, etc.;

– encode;

– select two or three prototypic urban applications for interoperability and/or cooperation;

– write local ontologies;

– complete the ontology if necessary.

The lessons learnt were the following:

– few people are interested by pre-consensus ontologies, that is by organizing and storing various definitions; they prefer to deal with a consensus already established; nevertheless the problem is "how do we reach to a consensus?";

– few people were interested by visual access, they prefer textual access likely because it is more familiar from them;

– several urban concepts do not exist everywhere (see Venice for an example for which a local vocabulary is used); the consequence was that using only English as a key-language was illusory.

6.4. Characteristics of geographic ontologies

A first definition is to claim that geographic ontologies are ontologies organizing geographic objects with conventional relations. In Figure 6.7, the beginning of such an ontology is given. However, can be immediately seen that such vision is totally insufficient to describe space.

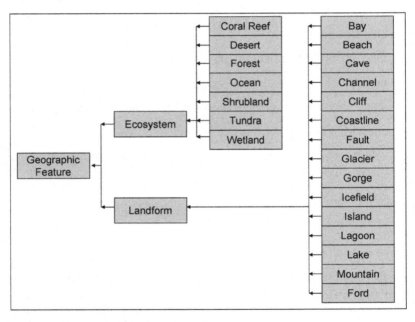

Figure 6.7. *Example of a geographic ontology only using* is_a *relations*

Another early example is taken from [SOW 09] in which a prototypic geographic ontology is described via geometric types of its features (Figure 6.8).

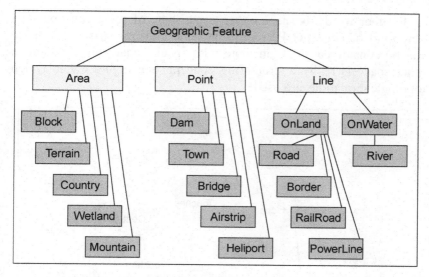

Figure 6.8. *Example of a naïve ontology [SOW 09]*
based on geometric types of features

As was previously explained, this is not acceptable because a lot of geographic features can have various geometric descriptions. Are "town", "bridge" really punctual objects? Maybe, this is right at some scales and at some scales; in other words, this ontology does not try to really model the geographic reality, but perhaps at cartographic level.

Indeed, different issues must be important in geographic ontology; the scope of this section will be to examine only four of them, namely the status of space, the spatial relations and linguistic problems.

6.4.1. *Space representation and management*

The first question to answer is the status of space since space can intervene in many descriptions. For instance, in ecology one has to describe biotope, in archaeology, the importance of places in which investigations were made, etc. But in applications such as in environmental and urban planning or in meteorology, space is truly a key-issue.

In other words, is space either an attribute of some concepts or a new kind of concept? Presently, in some existing ontologies, space can be considered as an attribute with special characteristics, but in other space is really a structuring concept. See Figure 6.9 in which topological relations are used.

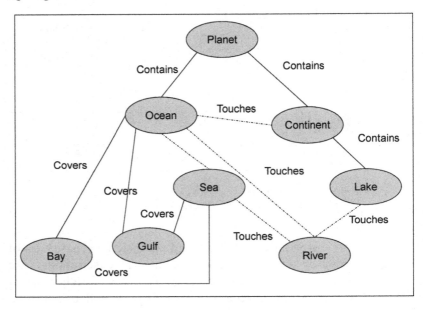

Figure 6.9. *Example of the beginning of a geographic ontology with spatial relations*

6.4.2. *Links with linguistics*

An important issue is language. As the scope of the European Towntology project was to design ontology for urban planning [TEL 07], the discussion was as follows: "do we either design a complete ontology in English, and then translate it into various other languages, or make several ontologies in different languages and then fusion them into an English ontology?". The dilemma was not really resolved in a satisfactory manner.

For instance, the well-known English term "bank" represents both a riverside and a financial institution. In other words, the first meaning will be translated in French and Spanish respectively, by *"rive"*, *"ribera"* and the second by *"banque"* and *"banco"*.

Let us examine a special case: in French language, the word *"quai"* defines a wharf, an embankment, a train platform or a street along a river (see Table 6.4 with English, Spanish and Italian translation). Moreover, in Spain, especially in Barcelona, *"rambla"* is a ravine or a special kind of broad avenue. In Venice, *"rioterà"* is a special type of pedestrian lane whereas other denominations are used such as *salizada, sottoportego, ramo, fondamenta, campiello, corte, calle, riva*, etc. As far as we know, those terms have no equivalent in English.

French	Picture	English	Spanish	Italian
Quai		Wharf	Muelle	Molo
		Riverside	Avenida a lo largo de un río	Lungofiume
		Platform	Andén	Binario

Table 6.4. *Different meanings of the French word "quai". Where to put it in an ontology written in French?*

As a consequence, different languages can have different concepts for describing features. In other words, two ontologies describing the same domain in different languages can be different. In international projects, this aspect can be difficult to solve. Furthermore, the same feature can have different names in different languages. Let us consider the river Danube. First in French, it is not considered a river but rather a "*fleuve*" which is defined as a river going to the sea. So, there is a topological relation between the river and the sea, a notion which is not integrated into English.

Gazetteers (see Chapter 8) were initially defined as dictionaries of place names (toponyms), but now, these are increasingly databases including not only the names of features but also their types and their geometric shapes. As a consequence, as an ontology is a knowledge resource organized by concepts and/or types, gazetteers are a knowledge resource based on geographic names.

6.5. Examples of geographic ontologies

A common method to easily develop and edit ontologies is to use Protégé[11], which was made by the Stanford University to create ontologies. In this paragraph, a few examples will be rapidly presented with Protégé. The first one is issued from the AKTiveSA[12] project, developed by [SMA 07] for humanitarian activities. Figure 6.10 gives an excerpt of their ontology in which some emphasis is given in various geographic bodies.

The second (Figure 6.11) is the beginning of an ontology which can be useful for remote sensing developed by [NEF 16]. The third (Figure 6.12) is taken from The European Urban Knowledge Network (EUKN[13]). The EUKN's main objective is to facilitate the interaction and exchange of valid and standardized knowledge across Europe on urban issues.

---

11 http://protege.stanford.edu/.
12 http://www.zaltys.net/ontology/AKTiveSAOntology.owl.
13 http://www.eukn.eu/.

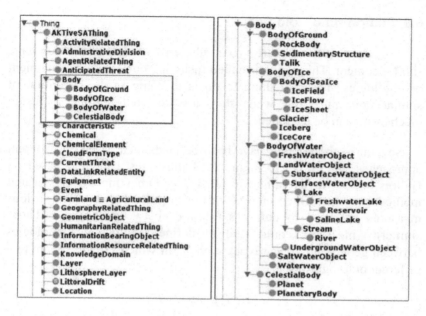

Figure 6.10. *Excerpt of the AKTiveSA ontology with emphasis on bodies*

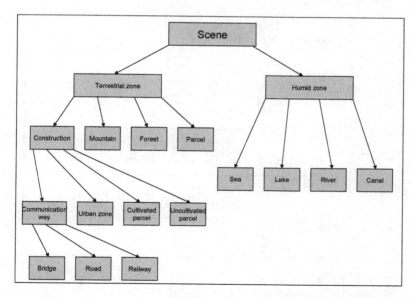

Figure 6.11. *Example of ontology for remote sensing [NEF 16]*

6.6. Fusioning ontologies

Sometimes it is of interest to combine two ontologies which have similar content. This action is called fusioning ontologies or alignment of ontologies. The first thing to do is to compare the definition of similar concepts. And when they are declared similar, a fusion mechanism can be launched.

Several methodologies have been defined, each based on different approaches. For instance, Figure 6.13 illustrates a method based on Gallois lattices as explained in [JEA 16]. Other more sophisticated methods were presented in [JAN 08] or in [NEF 16] in which various measures are used to define similarities between concepts. Those similarity measures could lead to define ontological homologies between geographic object types such as, in which Ω_1 and Ω_2 are two different ontologies:

$$\Omega_1\text{-}Type \bowtie \Omega_2\text{-}Type$$

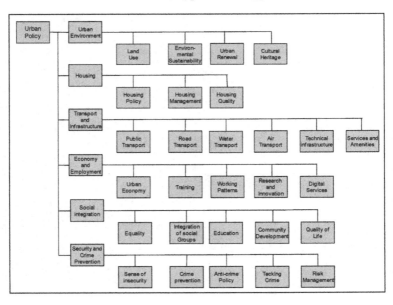

Figure 6.12. *The European Urban Knowledge Network (EUKN) top-level ontology. The thesaurus is composed of 254 concepts organized into five levels. From [TEL 07a]*

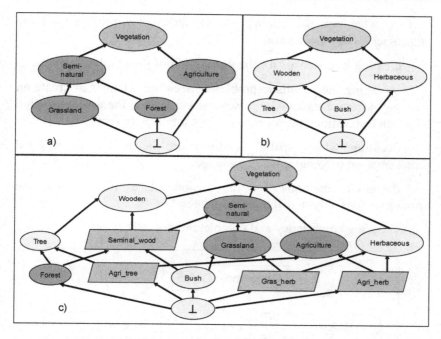

Figure 6.13. *Two ontologies a), b) and their fusion c). From [JEA 16]*

6.7. Conclusion and challenges regarding geographic ontologies

Finally, let me define a Geographic Ontology by stating that it is an ontology of geographic objects with geographic relations, especially spatial relations. All concepts can have specific attributes together with their data type (numeric, alphanumeric, string of characters, etc.).

Until now, ontologies and even geographic ontologies are only textual, but it could be interesting to consider and design multimedia ontologies, for instance by integration sounds (traffic noise, etc.) and pictures. An interesting case could be in image processing to recognize different types of land use for which a visual ontology could be very useful to help analyze satellite images and aerial photos. For this goal, an example of the beginning of a visual ontology is given in Figure 6.14.

But, what are the challenges we face concerning geographic ontologies? They are several:

– creating a consensus for the description of geographic features;

– by facing the linguistic problems, deciding whether to create an ontology in one language or in several languages; the actual solution based on translations from English is not totally satisfying;

– deciding which spatial relations are necessary for a good representation of geographic knowledge;

– designing methods for the fusion of existing geographic ontologies, possibly in different languages;

– checking consistency and completeness.

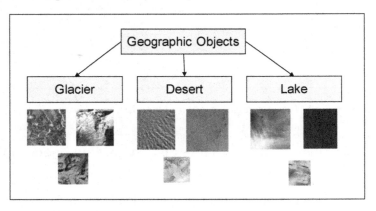

Figure 6.14. *Example of the beginning of a visual ontology for image processing of satellite images or aerial photos*

From an applicative point of view, the challenges could to be the following:

– what could be the role of geographic ontologies for geographic reasoning, for instance in environmental planning?

– how to use efficiently geographic ontologies for geographic information retrieval over Internet, for instance for tourism?

– what to include in ontologies? Could be of interest also to define attributes, constraints of concepts?

Now that geographic ontologies are defined and detailed, let us examine geographic structures as group of geographic objects linked not only by ontological relations but also by geographic relations.

7

Complex Geographic Objects and Structures

The goal of this chapter is to show that sometimes geographic objects are organized through very complex structures in which they are linked by various relations. In addition to simple collections, networks and tessellations are commonly found as complex structures. But look for instance at a city plan in which we can see that streets can have special organizations. By examining shape grammars some urban or territorial structures can emerge.

After briefly studying simple collections of geographic objects, networks and tessellations will be examined. Shape grammars will then be developed, in particular as organizing structures of cities.

7.1. Simple collections

Consider a woodland: it is an area in which there are trees. In other words, woods or forests can be considered as a collection of trees. If we only consider trees, the wood is a set of trees, that is a set of points or small areas, or more exactly from a geometric point of view, the wood can be considered as the convex hull (section 5.1.5) of trees $(CH(t_i))$ only if they are numerous (for instance greater than 50). So, the following rule (Rule 7.1) can hold:

$\forall t_i \in GO$, $\Omega\text{-}Type(t_i) = $ "Tree", $Concept$ ("Wood") $= Has\_a$ ($Concept$("Tree")), $\exists Z \in GO$: $Card(t_i) > 50$ $\Rightarrow$ $\{\Omega\text{-}Type(Z) = $ "Wood"; $G\text{-}Type\ (Z) = Area$; $Geom(Z) = CH(t_i)\}$	Rule 7.1

Another definition can be based on to granularity of interest, those trees, as areas, can coalesce to give a bigger area, that is wood. See Figure 7.1 for an example. The rule (Rule 7.2) will be as follows for instance by assigning a 2 m buffer zone to each tree:

$\forall t_i \in GO$, $\Omega\text{-}Type(t_i) = $ "Tree", $Concept$ ("Wood") $= Has\_a(Concept$("Tree")), $\exists Z \in GO$ $\Rightarrow$ $\{\Omega\text{-}Type(Z) = $ "Wood"; $G\text{-}Type\ (Z) = Area$; $Geom(Z) = Union(Buffer(t_i, 2))\}$	Rule 7.2

Let us now consider a set of city-blocks. At a certain scale, we have a set of areas. But when the scale becomes smaller, the streets between the city-blocks can reach the threshold of granularity of interest. So, the city-blocks will coalesce to give a bigger zone, often named a quarter. By continuing, all those city-blocks will coalesce to give a city (Figure 7.2). The rule (Rule 7.3) can be written:

$\forall c_i \in GO$, $\Omega\text{-}Type(c_i) = $ "City-Block", $Concept$ ("City") $= Has\_a$ ($Concept$("City-Block")), $\exists Z \in GO$ $\Rightarrow$ $\{\Omega\text{-}Type(Z) = $ "City"; $G\text{-}Type\ (Z) = Area$; $Geom(Z) = CH(c_i)\}$	Rule 7.3

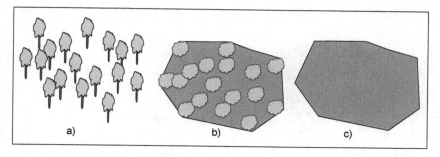

Figure 7.1. *Example of a collection of trees represented by small areas:*
a) a collection of trees; b) with their convex hull; c) the convex hull
representing the forest

In the same spirit, consider geographic objects made of several smaller objects, such as airports, universities, hospitals, barracks, big plants, etc. At some scale, different constructions can be seen; but as the scale diminishes, neighboring objects coalesce similarly. Then they become a small area, then a point and finally void.

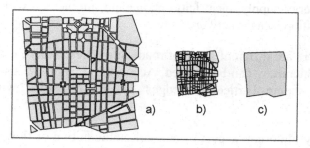

Figure 7.2. *Example of a set of city-blocks coalescing into an area:*
a) city-blocks; b) city-blocks at a smaller scale in which streets are
no more visible; c) coalescing into a single area

7.2. Ribbon graphs and networks

Now, in terms of sewerage, water supply, electricity networks, hydrological and road networks, they all can be considered as ribbon networks (Figure 7.3).

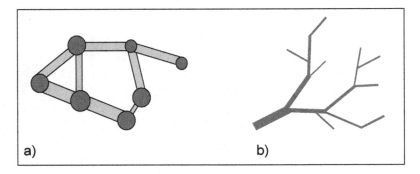

Figure 7.3. *Ribbon networks: a) theoretical structure;
b) example of a river network modeled as ribbon network*

A graph is usually defined as an ordered pair $G = (V, E)$ comprising a set V of vertices or nodes together with a set E of edges linking two vertices. In our case, a ribbon network is a graph in which edges are ribbons or chain of ribbons, and nodes, areas. When modeling by means of ribbon networks in a city, let us mention water, gas or energy supply, electricity, phone and Internet networks, public transportation, sensor networks, etc.

As there are different kinds of roads, turnpikes, streets, etc, seldom a sort of hierarchy can be defined. According to applications, in those networks, some vertices or edges can be discarded with some thresholds.

A very important aspect of network is the notion of flows. A flow can be defined as an attribute of edges. The more important flows concern population (for instance home-to-work journeys) and freight, and the great majority of studies regarding transportation is based on flows. In Figure 7.4, several types of flows are depicted; Figure 7.4(a) represents a unidirectional flow; Figure 7.4(b), a bidirectional flow; Figure 7.4(c), a diverging flow, for instance to represent out-migrations; and Figure 7.4(d), a converging flow, for instance in-migrations.

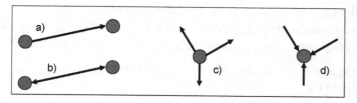

Figure 7.4. *Several types of flows; a) unidirectional flow b) bidirectional flow; c) diverging flow; d) converging flow*

From a formal point of view, the following type must hold *Network_of* (*Edges, Nodes*) together with G-*Type* (*Edges*) = *Ribbon* and G-*Type* (*Nodes*) = *Area*. Remember that some ribbons must be perhaps cut into elementary pieces. As examples let us mention:

G-*Type* (*Metro_Network*) = *Network_of* (*Metroline_segments, Metro-stations*)

G-*Type* (*Route_Network*) = *Network_of* (*Road_segments, Crossroads*).

G-*Type* (*Hydro_Network*) = *Network_of* (*River_segments, Confluence_Areas*).

Consider now the various networks for transportation, roads, street-cars, buses, bikes, railways, metros, etc. For interconnecting those networks, some multimodal platforms are usually constructed: at the vicinity of railway stations, usually there are bus stops; usually metro stations are near crossroads, etc. Eventually, some network nodes are connected to other network nodes. Let's call them hyper-nodes. They will be used to find multimode travel from a point *A* to a point *B*.

It is important to mention that several procedures are needed to manipulate networks, such as adding an edge or a vertex, and amalgamating two networks.

7.3. Tessellations

For instance, the conterminous States in the USA form a tessellation. Generally speaking, administrative subdivisions form tessellations, sometimes hierarchical tessellations. Let us consider a domain D and several polygons (sometimes called cells) which are not necessarily connected, C_i; they form a tessellation (Figure 7.5(a)) if:

– for any point p_k, if p_k belongs to D then there exists C_j, so that p_k belongs to C_j;

– for any p_k belonging to C_j, then p_k belongs to D.

Note that:

– a tessellation can also be described by Egenhofer relations applied to C_i and D;

– in practical cases, due to measurement errors, this definition must be relaxed in order to include sliver polygons; let's call them "loose tessellations". See Figure 7.5(b). The good-looking tessellations are also called valid tessellations;

– one must include spherical tessellations and 3D tessellations;

– for terrain modeling, a TIN (Triangulated irregular network) constitutes a special case of 3D triangular tessellation;

– sometimes one can have a valid tessellation from an administrative point of view, but not valid from a mathematical point of view; for instance a file of regions in France with measurement errors generating seldom the so-called sliver polygons by small overlaps and voids between cells.

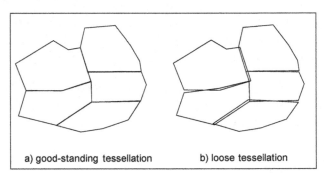

a) good-standing tessellation   b) loose tessellation

Figure 7.5. *Examples of irregular tessellations. a) a mathematical good-looking tessellation (valid); b) a practical tessellation (loose tessellation) with sliver polygons in which errors are voluntarily exaggerated*

By defining integrity constraints, it is easy to state whether a so-called tessellation is valid or not ([LAU 02] and [SER 00]). But when one is dealing with a loose tessellation, the transformation into a valid tessellation is a much more complex task.

From a formal point of view, tessellations must be considered as a new kind of geometric type, a sort of additional modifier. More exactly, it will be loose tessellation in order to get rid of sliver polygons. For instance, considering the USA as a tessellation of States, we can write:

G-*Type* (*State$_i$*) = *Area*

G-*Type* ("USA") = (*Area, Tessellation_of* (*State$_i$*)).

Several procedures are needed to manipulate tessellations. The first one is to transform loose tessellations into valid ones. Others may be of importance such as creating tessellations by means of spatial analysis such as gerrymandering for elections. See [LAU 93], for instance, for details.

7.3.1. *Hierarchical tessellations*

Consider territorial divisions. In some countries, there are states, provinces, regions, municipalities, etc. So, these divisions can be considered tessellations of tessellations or moreover hierarchical tessellations. According to the level of reasoning, lower subdivisions can be discarded.

7.3.2. *Reduction of tessellations*

In any tessellation, there exist cells of various sizes, some of which can be very small. According to the granularity of interest, the smaller cells will disappear. For instance, consider Europe and some small countries such as Vatican, Monaco, Andorra and Luxemburg. Their disappearance or absorption can follow several rules (Figure 7.6):

– Vatican-style: when a cell is part of another one (*Contains*), the smaller cell disappears (see Rule 7.4);

– Monaco-style: when a cell is located at the border of a bigger cell (*Covers*), the smaller cell is integrated into the bigger one;

– Andorra-style: when a cell is located at the border of two bigger cells, the borders of the bigger cells are reorganized, for instance fifty-fifty;

– Luxembourg-style: when a cell is located at the border of three or more bigger cells, the borders of the bigger cells are reorganized accordingly.

From-to mutation	Initial scale	Smaller scale
Vatican-style	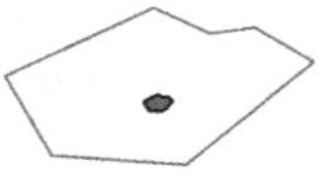	
Monaco-style		
Andorra-style	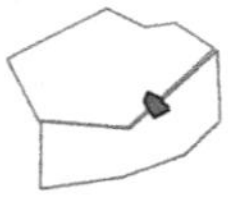	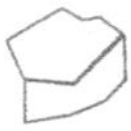
Luxemburg-style		

Figure 7.6. *Rules for smaller cell absorption*

$\forall O \in GO, \forall t^i \in GO, G\text{-}type(t^i) = Area,$ $G\text{-}Type(O) = (Area, Tessellation\_of(t^i)),$ $Contains(O, t), t_\sigma \equiv 2Dmap(t, \sigma):$ $Area(t_\sigma) < \varepsilon$ $\Rightarrow$ $G\text{-}Type(t_\sigma) = void$	Rule 7.4

Now that common geographic structures are defined, let us present some more complex geographic structures derived from shape grammars.

7.4. Shape grammars and applications to geographic objects

Look for example at buildings. Generally the organization of floors is the same; in other words, they derive from, or are produced by a shape grammar. A lot of manmade objects show repetitions as illustrated in Figure 7.7.

The goal of this section is to give some information regarding shape grammars as a very sophisticated way to define relations between spatial objects. After a rapid introduction and some examples taken from various domains and especially in architecture, some other applications in town, community and landscape planning will be developed.

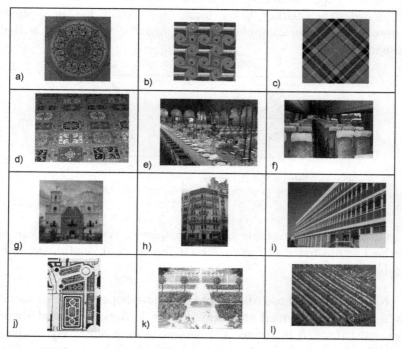

Figure 7.7. *Various examples of applied shape grammars: a) for decoration; b) pattern of a tabric; c) a tartan; d) pavement; e) a table for a dinner; f) inside a bus; g) a church; h) and i) various façades; j) and k) french-style gardens; l) terraced-houses*

7.4.1. *Introduction to shape grammars*

Starting from one of several initial shapes (squares, triangles, cubes, spheres, etc.), a sequence of shapes can be generated through grammar. Hence, shape grammars allow us to define spatial objects with spatial relationships derived from grammar.

According to Stiny [STI 80], a shape grammar consists of shape rules and a generation engine that selects and processes rules. A shape rule defines how an existing (part of a) shape can be transformed. A shape rule[1] consists of two parts separated by an arrow pointing from left to right. The part left of the arrow is termed the Left-Hand Side (LHS). It depicts a condition in terms of a shape and a marker. The part right of the arrow is termed the Right-Hand Side (RHS). It depicts how the LHS shape should be transformed and where the marker is positioned. The marker helps locate and orient the new shape. Figure 7.8 gives an example of shapes together with a marker (half hexagon).

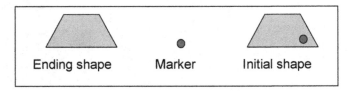

Figure 7.8. *Example of components of a shape grammar (half hexagon example)*

Continuing the example of half hexagons, several rules can be defined as illustrated in Figure 7.9. Of course, additional rules can be designed.

Now, based on those rules, more complex shapes can be generated by a set of rules named programs. Figure 7.10 gives several shapes generated through programs. For instance, by successively applying the rules (3, 1, 3, 1, 3, 4), one gets Figure 7.10(a); by means of the

---

1 Remark that for shape grammars, one deals with repetitive rules whereas for fractal geometry, one deals with recursive rules.

program (1, 1, 2, 2, 2, 4), the shape given in Figure 7.10(b) is generated; and by program (2, 2, 2, 2, 2, 4), Figure 7.10(c).

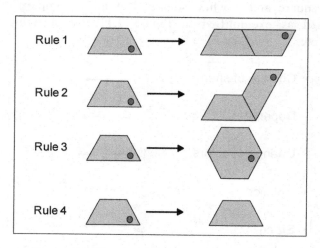

Figure 7.9. *Examples of rules for half hexagons*

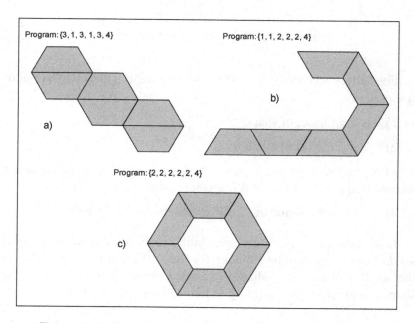

Figure 7.10. *Examples of shapes generated through programs*

The previous cases were defined by using a single initial shape. For instance, to define a chessboard, two initial shapes must be used, a black square and a white square. But more complex rules can be defined as exemplified in Figure 7.11, by changing shapes, rotation, etc.

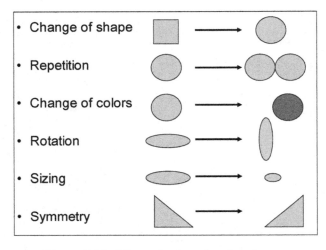

Figure 7.11. *Other rules transforming shapes*

Formally, according to [STI 80], a shape grammar has four components:

1) S is a finite set of shapes;

2) L is a finite set of symbols;

3) R is a finite set of shape rules of the form $\alpha \rightarrow \beta$, where α is a labeled shape in $(S, L)^+$, and β is a labeled shape in $(S, L)^*$; and

4) $/$ is a labeled shape in $(S, L)^+$ called the initial shape.

In a shape grammar, the shapes in the set S and the symbols in the set L provide the building blocks for the definition of shape rules in the set R and the initial shape $/$. Labeled shapes generated using the shape grammar are also built up in terms of these primitive elements.

Historically speaking, the Italian architect Andrea Palladio (1508–1580) invented a way to organize rooms in a building. His method was then called "Palladian grammar" by Stiny and Mitchell [STI 78]. An example is given Figure 7.12 in which one can find some rules for organizing rooms in a building.

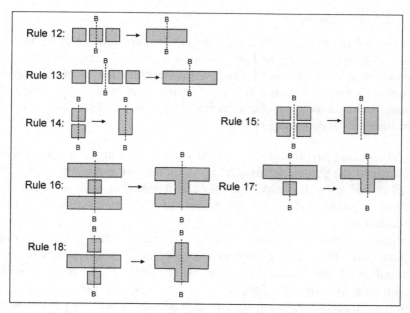

Figure 7.12. *Example of Palladian grammar taken from [STI 78]*

To conclude this section, let me briefly say that shape grammars can define a class of very sophisticated spatial relations between shapes and spatial objects. The application in architecture and in garden organization can be seen as a first step for landscape and town planning.

An often-mentioned drawback is called visual monotony, that is repetitions can be seen annoying for several persons. However, a sense of unity, of harmony can be generated through shape grammars when they are elegantly used.

7.4.2. Applications to landscape and town planning

In archaeology, evidence shows that the first excavated cities present some regular patterns. Since then, urban planners have tried to use patterns to organize cities and territories.

7.4.2.1. Roman epoch

Even if in the past the present concept of shape grammars for urban and landscape planning had not yet been created, allusions to it can be found in several places. Indeed we have evidence from the early Roman Republic and the Roman Empire. Surely, the skills behind those realizations came from ancient Egyptians or Etruscans. Look, for instance, at centuriation.

According to *Wikipedia*[2], Roman centuriation is a geometric schema of the parcels forming a city or an agricultural territory, which was traced with the help of a line and a team, in each new colony where Romans settled. They began to use centuriation for the foundation, in the fourth century BCE, of new colonies in the ager Sabinus, northeast of Rome. The development of the geometric and operational characteristics that were to become standard came with the founding of the Roman colonies in the Po valley, starting with Ariminum (Rimini) in 268 BCE. There were several patterns and variety of accommodation measures. The most common pattern was the *ager centuriatus*. After choosing the City Centre (*umbilicus*), the surveyor of the time would have traced two road axes perpendicular to each other from it: the first in East-West direction, called *decumanus maximus*, the second in a North-South direction, called *cardo maximus*. After having bounded the city, these two streets would extend into the surrounding agricultural territory through the gates of the city walls.

In the Venetian area, the Roman centuriation is still visible in the countryside and is better known as the *graticolato romano* (Figure 7.13).

---

2 https://en.wikipedia.org/wiki/Centuriation.

Figure 7.13. *Traces of centuriation in Northern Italy called graticolato romano. Source: http://pre-evolutionary2.rssing.com/chan-1016278/all_p220.html*

It has been suggested that the Roman centuriation system inspired Thomas Jefferson's proposal to create a grid of townships for survey purposes, which ultimately led to the United States Public Land Survey System. The similarity of the two systems is empirically obvious.

7.4.2.2. Urbanization

The first urban planner was most likely the Greek Hippodamos of Milet (498–408 BC) who was in charge of re-planning the harbor after a war. Of course, other individuals had thought about cities before him, but apparently, he was the first to give a rationale for organizing cities according to a grid plan. During the Roman Empire, several cities were founded with more or less the same gridded structure.

In Figure 7.14, several maps of cities are given emphasizing the evolution of the design from Hippodamos of Milet (Figure 7.14(a)), the city of Palmanova in Italy (Figure 7.14(b)), the city of La Plata in

Argentina (Figure 7.14(c)), the city of Brasila in Brazil (Figure 7.14(d)), the city of Erice in Sicily (Figure 7.14(e)) and Beijing's ringroads (Figure 7.14(f)). It is easy to see that each of them corresponds to particular spatial patterns.

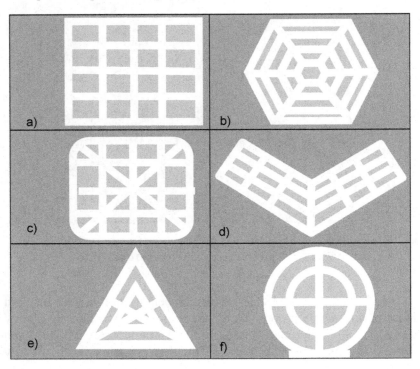

Figure 7.14. *Illustration of spatial patterns in cities: a) plan designed by Hippodamos of Milet; b) Palmanova; c) La Plata; d) Brasilia; e) Erice in Sicily; f) Beijing's ringroads*

7.4.2.3. *Small community planning*

Except perhaps Brasilia, nowadays few cities have been designed with regular patterns. However, grammar shapes are hidden in smaller communities. Take for instance terraced-houses in the UK, and gated communities as exemplified in Figure 7.15.

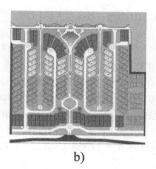

a)   b)

Figure 7.15. *Small communities planned with regular patterns or shape grammars: a) terraced-houses; b) a gated community*

In addition, as explained by [HAL 08], shape grammars can also be used in urban planning especially at 3D levels. It is important to mention that for some people, this kind of spatial order can be both elegant and sometimes visually boring.

7.5. Complex geographic objects and their relations

As mentioned earlier, many geographic features are complex, especially when manmade: consider for example airports, universities, schools, factories and barracks. They are all composed of several buildings, parking lots, roads, etc. Consider also engineering networks: in addition to power lines, electric networks include poles, transformers, etc. Similarly sewerages include tubes, manholes and water treatment plants.

They all have in common to be seen either from a set-theoretical approach (*Part_whole* relations) or from a topological point of view (*Contains* relation).

Generally speaking, engineering networks are set along streets. If we consider for instance water supply and electricity supply network, a geometric homology can be set:

Water_Supply_Network ₪ *Electricity_Network*

Water_Supply_Network ₪ *Sewerage_Network.*

7.6. Conclusion

As explained in this chapter, geographic objects and especially man-made geographic objects are organized into structures such as ribbon networks, tessellations or deriving from shape grammars. Eventually, new geometric types must be considered. In section 4.9, geometric types were defined as:

G-*Type* = {*Point, Line, Area, Ribbon, Void, Null*}

Now, bearing all that in mind, we must extend this definition by integrating tessellations (of areas) and networks (of ribbons) by remembering that ribbon networks supersede line networks or graphs:

G-*Type* = {*Point, Line, Area, Ribbon, Tessellation_of, Network_of, Void, Null*}

But one of the problems concerns the identity of geographic objects and structures. It will be the scope of the next chapter to explore the concepts of gazetteers which initially were simple lists of place names, but nowadays they more often tend to be databases and knowledge bases.

8

Gazetteers and Multilingualism

By definition, a gazetteer is a directory of place names or toponyms [FUB 12, KES 09]. However, now, more and more gazetteers are becoming complex databases. Since they increasingly include other attributes of the named features, they tend to become toponym ontologies [JAK 11, HEC 13]. However, there is also the problem of languages; we touched on this briefly in the chapter dealing with ontologies but since a lot of places have different names in different languages, this problem is crucial[1].

Beneath a geographic name, various objects or features can exist. On the Earth, only a few real points have names, such as the North and South Poles and some mountain summits, and a few lines, such as the Equator, the Tropic of Cancer, the Tropic of Capricorn, the Greenwich Meridian, the Polar Circle, etc. The majority of names is given to areas, since even rivers are areas or may be modeled as lines or ribbons. As previously mentioned, they must be considered as simply connected (with islands and holes), and they can be replaced by their centroids for some operations. In some geographic databases, for instance, the geographic object named "Italy" can include the Vatican and San Marino, whereas those places do not belong officially to the country named Italy.

---

1 A preliminary version was published in [LAU 15a].

8.1. Generalities

Indeed, in addition to a pure list of place names, it is necessary to locate geographic objects with accuracy and assign them a ontological type. Moreover, a place can have different names in different languages and different periods of time (see Prolegomenon #10). Let us first examine a few well-known examples:

– "Mississippi" can be the name of a river or of a state;

– the city, "Venice", Italy, is also known as "Venezia", "Venise", "Venedig", respectively, in Italian, French and German;

– the local name of the Greek city of "Athens" is "Αθήνα"; read [a'θina];

– "Istanbul" was known as "Byzantium" and "Constantinople" in the past;

– the modern city of Rome is much bigger than in Romulus's time;

– there are two Georgias, one in the United States and another one in Caucasia;

– the toponym "Milano" can correspond to the city of Milano or the province of Milano;

– the river "Danube" crosses several European countries; in practically each country, it has a different name, "Donau" in Germany and Austria, "Dunaj" in Slovakia, "Duna" in Hungary, "Dunav" in Croatia and Serbia, "Dunav" and "Дунав" in Bulgaria, "Dunărea" in Romania and in Moldova and "Dunaj" and Дунай" in the Ukraine. It is also called "Danubio" in Italian and Spanish, "Tonava" in Finnish and "Δούναβης" in Greek. Moreover, its name is feminine in German and masculine in some other languages;

– sometimes, names of places can be also names of something else; for instance "Washington" can also refer to George Washington or anybody with this first name or last name;

– In the U.K., there are several rivers named Avon;

– some place names are formed of two or more words; for instance, "New Orleans", "Los Angeles", "Antigua and Barbuda", "Trinidad and

Tobago", "Great Britain", "Northern Ireland", "Tierra del Fuego", "El Puente de Alcántara", etc.;

– some very long names can have simplifications; the well-known Welsh town "Llanfairpwllgwyngyllgogerychwyrndrobwllllantysiliogogo goch" is often simplified in "Llanfair PG" or "Llanfairpwll";

– some abbreviations can be common, such as "L.A." for "Los Angeles", whereas its name at its inception was "El Pueblo de Nuestra Señora la Reina de los Ángeles del Río de la Porciúncula";

– Peking became Beijing after a change of transcription to the Roman alphabet; but the capital of China has not modified its name in Chinese;

– in some languages, grammatical gender is important, so that place names can be feminine or masculine; for instance, in French, Italian and Spanish, names such as "Japan", "Brazil" and "Portugal" are masculine, whereas "Argentina", "Bolivia" and "Tunisia" are feminine;

– in addition, as the great majority of toponyms are singular, some can be plural, like "The Alps"; but for "The Netherlands", the situation is more complex: plural in French (Les Pays-Bas), in Italian (I Paesi Bassi) and in Spanish (Los Países Bajos), whereas singular and plural are both acceptable in English (The Netherlands are, The Netherlands is);

– as the toponyms in several languages look similar (Venice/Venezia), others are very different. For instance in Belgium, the city of Mons (in French) is known as Bergen in Flemish. Not far from this city, the French city of Lille is named Rijsel in Flemish;

– some places have nicknames; for example Dixieland, Big Apple, City of the Lights, etc.;

– do not forget that in some languages, toponyms can have declensions, for instance for the Rhine River in German (der Rhein, des Rheins, etc.).

Consider now the toponym "Granada": there are places in practically all Spanish-speaking countries bearing this name:

– a small country located in the Caribbean Islands;

– in Spain, the capital city of the eponymous province, a few other places located in Barcelona and Huelva provinces and a river in the Vizcaya province;

– in Colombia, three cities with this toponym;

– in the U.S., cities in California, Colorado, Kansas, Minnesota, Mississippi, etc.;

– in Mexico, a city in Yucatán;

– in Nicaragua, a city capital of the eponymous department;

– and in Peru, a district.

As a consequence, there is a very complex many-to-many relationship between places and place names (Figure 8.1).

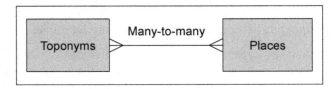

Figure 8.1. *Very complex many-to-many relations link places and their names*

Among place names, there are street names together with the number in the street (civic number); these are not so easy to handle. This is very important, not only for the automatic processing of postal addresses, but also for all applications connected to an emergency. The Urban and Regional Information Systems Association (URISA)[2] association has organized many conferences on the topic. The specificities of street names are as follows:

– some streets are of a few dozen yards long, whereas others several miles;

– in some human settlements, streets have no names;

---

2 http://www.urisa.org.

– sometimes, there are variations about the way to write some street names; for instance "3rd Street", "Third Street", "Third St"; the words "avenues" and "boulevards" are commonly simplified into "Ave" and even "Blvd" or "Bd";

– in some countries, especially in Spanish-speaking countries, the equivalent of the words "street", "avenue", etc., are usually removed;

– in some places, streets can have several names; for instance, in New York City, "Sixth Avenue" is also known as "Avenue of the Americas".

As a main consequence, the name of a place cannot be a unique ID from a computing point of view. In order to clarify, let us give a few definitions:

– toponym is the general name of a geographic feature;

– endonym is a local name in the official language of the country or in a well-established language occurring in that area where the feature is located; there may be several toponyms in countries with different official languages (Brussel in Flemish, Bruxelles in French);

– exonym is a name in languages other than the official languages; for instance Brussels in English;

– archeonym is a name that existed in the past: for instance, Byzantium for Istanbul;

– hyperonym and hyponym are the names of places with a hierarchy; hyponym is the opposite of hyperonym; for instance, Europe is a hyperonym of France, whereas France is a hyponym of Europe;

– meronym is a name of a part of a place without a hierarchy; sometimes the expression partonym is used; for instance "Adriatic Sea" is a meronym of the Mediterranean Sea;

– hydronym is a name of a waterbody;

– oronym is a name for a hill or a mountain.

Figure 8.2 gives the essential elements of a gazetteer, the names, the features, the dates and everything regarding geometry and georeferencing according to [JAK 11].

In addition, places, such as airports, can have several names. Sometimes, their International Air Transport Association (IATA)[3] codes are used: the well-known New York City airport, John F. Kennedy International Airport, is often referred to as JFK. Zip codes or postcodes can also be considered as toponyms. However, the definition of postcodes differs according to country: in some cases, one postcode can correspond to a few houses, and in others some hundred thousand inhabitants.

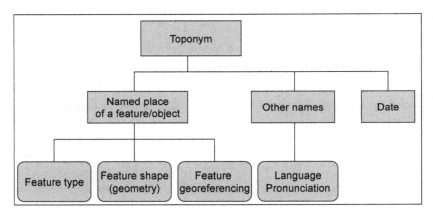

Figure 8.2. *Essential elements of a toponym, according to [JAK 11]*

To conclude this section, in an automatic system for searching geographic information in the web (often known as GIR, geographic information retrieval), a preliminary phase of disambiguation is necessary, since the name can correspond to something that is not geographic (Figure 8.3). For more details, please refer to [SAL 13].

---

3 http://www.iata.org.

Let us define as a literal a string of characters (perhaps including blank spaces, hyphens and numbers): this literal may be a toponym, the name of a person (Washington) or something else (China and porcelain). Toponyms can be described as literals.

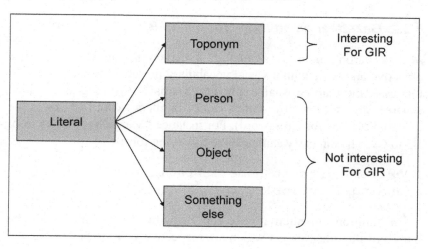

Figure 8.3. *Disambiguation of literals to extract toponyms. GIR meaning geographic information retrieval*

8.2. Examples

Generally speaking, a gazetteer is designed for a specific activity, for instance to help post offices, to assist the history of a region, etc. As a consequence, several gazetteers can have different structures. Let us examine a few examples.

8.2.1. *Simple gazetteer*

A simple gazetteer consists of a list of place names and equivalence relationships between them, such as:

"U.S.A." ≡ "United States of America".

Therefore, the database tables are as follows:

```
Placenames (ID, toponym);
Equivalence (ID1, ID2).
```

8.2.2. Gazetteer as an index for a map (street directory)

The starting point is a map of a certain region with a precise objective and scale with a visual vocabulary presented in the legend. In this case, the map is usually split into a crossword-like grid in which squares are located by letters and numbers identified by CW-location (CW for Crossword). For instance "Main Street" goes from B3 to C7. The directory can have the following forms:

```
Location1 (street-name, CWbeginning-location,
CWending-location)
```

In addition, an alternative could be with street names with the names of the other streets which are respectively at the beginning and at the end of the road.

```
Location2 (street-name, beginning-street-name,
ending-street-name)
```

8.2.3. Gazetteer for a local post-office

For the post-office, the gazetteer can have the previous forms, but in addition, it can also include several important monuments, administrations and enterprises that can be stored:

```
Urban-feature (name, street-address, postcode)
```

8.2.4. Gazetteer for hydrology

Here, there are only names of rivers, lakes, seas, waterways, etc. Important relations are for tributaries and possible estuaries with the sea in which id, id1 and id2 are computer object identifiers or access-keys.

```
Hydronym (id, onto-type, geometry)
Endonym (id, hydronym)
Exonym (id, language, hydronym)
Tributary (id1, id2, location)
Estuary (id1, id2, location)
Meronym (id1, id2).
```

8.2.5. Gazetteer for the history of a place

Here, we essentially deal with ancient names. Let us start with the actual toponyms.

```
Placename (id, onto-type, geometry, beginning-
date)
Archeonym (id, language, toponym, geometry,
beginning-date, ending-date)
Exonym (id, language, toponym).
```

8.2.6. Gazetteer covering several countries

```
Placename (id, onto-type, geometry, beginning-
date)
Exonym (id, language, toponym)
Hydronym (id, onto-type, geometry)
Endonym (id, hydronym)
Exonym (id, language, hydronym)
Meronym (id1, id2).
```

8.3. Existing systems

Concerning ontologies and gazetteers, several systems exist. Let us rapidly present two of them, GeoNames[4] and GeoSPARQL[5].

---

4 http://www.geonames.org.
5 http://geosparql.org/.

8.3.1. *GeoNames*

The GeoNames database contains over 10,000,000 geographic names corresponding to over 7,500,000 unique features. All features are categorized into one out of nine feature classes and further subcategorized into one out of 645 feature codes. Beyond names of places in various languages, the data stored include latitude, longitude, elevation, population, administrative subdivisions and postal codes. Among spatial relationships, GeoNames utilizes a special way to model hierarchical tessellations:

– children, that is the list of administrative divisions (first relative sublevel);

– hierarchy, that is the list of toponyms higher up in the hierarchy of a place name;

– contains, that is the list of all features within the feature;

– siblings, that is the list of all siblings of a toponym at the same level.

For instance, here is an excerpt of the description of Sicily in which the tag `<ToponymName>` corresponds to an endonym and `<name>` to an exonym; the number 9 of children corresponds to the nine provinces of Sicily:

```
<geoname>
    <toponymName>Sicilia</toponymName>
    <name>Sicily</name>
    <lat>37.75</lat><lng>14.25</lng>
    <geonameId>2523119</geonameId>
    <countryCode>IT</countryCode>
    <countryName>Italy</countryName>
    <numberOfChildren>9</numberOfChildren>
</geoname>
```

8.3.2. *GeoSPARQL*

GeoSPARQL is a standard for the representation and querying of geospatially-linked data for the Semantic Web from the Open Geospatial Consortium (OGC)[6]. It can be seen as an extension of SPARQL[7]. The definition of a small ontology based on well-understood OGC standards is intended to provide a standardized exchange basis for geospatial Resource Description Framework (RDF) data (see section 2.3) which can support both quantitative and qualitative spatial reasoning and querying with the SPARQL database query language.

However, with SPARQL, some simple geographic queries, that is, without geometric information and spatial relationships, can be launched. For instance: "What are all of the country capitals in Africa?":

```
PREFIX abc: <http://example.com/exampleOntology#>
SELECT ?capital ?country
WHERE {
      ?x abc:cityname ?capital;
      abc:isCapitalOf ?y.
      ?y abc:countryname ?country;
      abc:isInContinent abc:Africa.
      }
```

However, with GeoSPARQL, not only geometric attributes (shapes), but also Egenhofer/RCC (Confer section 5.2.1) topological relations can be invoked.

In addition, the following geoprocessing functions are integrated: distance, buffer, convex hull, intersection, union, difference, etc. The general structure and an example are given in Figure 8.4, in which WKT means "well known text" as defined by OGC. To get the Washington Monument, one has to write a small filter as a minimum bounding rectangle (MBR) as exemplified in the GeoSPARQL user

---

6 http://www.opengeospatial.org.
7 http://www.w3.org/2009/sparql/wiki/Main_Page/.

guide (by using an MBR, the search space is reduced in order not to run the query against the whole database):

```
PREFIX geo:
<http://www.opengis.net/ont/geosparql#>
PREFIX geof:
<http://www.opengis.net/def/function/geosparql/>
PREFIX sf: <http://www.opengis.net/ont/sf#>
PREFIX ex:
<http://example.org/PointOfInterest#>
SELECT ?a
WHERE {
      ?a geo:hasGeometry
      ?ageo.
      ?ageo geo:asWKT
      ?alit
      FILTER( geof:sfWithin(?alit, "Polygon
      ((-77.089005 38.913574,-77.029953
      38.913574,-77.029953 38.886321,
      -77.08900538.886321,-77.089005
      38.913574))"^^sf:wktLiteral))
      }
```

For instance, a query for getting the airports near London is as follows:

```
   PREFIX co:
<http://www.geonames.org/countries/#>
   PREFIX xsd:
<http://www.w3.org/2001/XMLSchema#>
   PREFIX geo:
<http://www.w3.org/2003/01/geo/wgs84_pos#>
   SELECT ?link ?name ?lat ?lon
   WHERE {
      ?link gs:within(51.139725 -0.895386
      51.833232 0.645447).
      ?link gn:name ?name.
      ?link gn:featureCode gn:S.AIRP.
      ?link geo:lat ?lat.
      ?link geo:long ?lon
      }
```

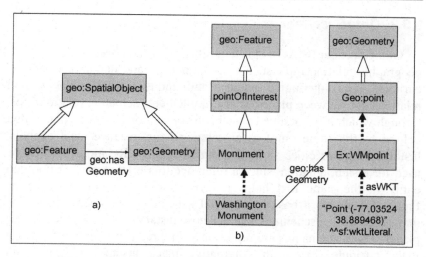

Figure 8.4. *Example of describing geographic entities in GeoSPARQL: a) generic structure; b) example for a monument*

If somebody is looking for land parcels with some type of commercial zoning that touches some arterial street, the query is the following:

```
SELECT ?parcel ?hwy
WHERE {
        ?parcel rdf:type :Commercial.
        ?parcel rdf:type ogc:GeometryObject.
        ?hwy rdf:type :Arterial_Street.
        ?hwy rdf:type ogc:GeometryObject.
        ?parcel ogc:touches ?hwy
        }
```

Now that the notions of gazetteers and geographic ontologies in multiple languages have been clarified, let us work with these elements to enrich them.

8.3.3. OntoGazetteer

OntoGazetteer [MAC 11] is a gazetteer that includes topological and geographic relationships, such as contained by and neighbor to, and is enriched with alternative names, ambiguously named places, and relationships between places. It also implements semantic relationships, through which it is possible to establish connections between places that belong to the same ontological category, for instance state capitals, historical sites or cities along the same highway. Relationships can be used for disambiguation, since a reference to a related geo- or non-geo entity in the same text helps inferring which the correct place is. Figure 8.5 shows a basic conceptual schema for OntoGazetteer. The main class of this schema is Place, whose instances represent real world places. The place names are represented in three different ways: 1) as a Place attribute, 2) as an Alternative place instance or 3) as an Ambiguous name instance. These classes keep alternative names for the same place, and a list with identifiers of homonymous places, respectively. Although this schema has been successfully implemented as a relational database, searching for a name implies looking into three different structures (Place, Alternative place and Ambiguous name), making it harder to detect and solve ambiguities.

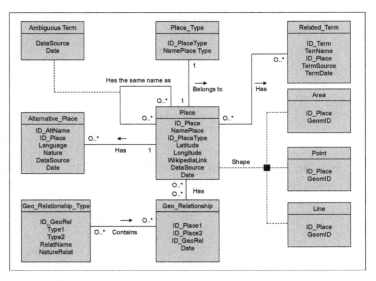

Figure 8.5. *OntoGazetteer conceptual schema from [MAC 11]*

8.3.4. *Metagazetteer*

In [MOU 16], the authors have proposed an extension of the previous system by creating a metagazetteer. A metagazetteer can be defined as a mediation framework to access and integrate distributed gazetteer resources. They propose integrating linked data sources to create a gazetteer that combines a broad coverage of places with urban detail, including content on geographic and semantic relationships involving places, their multiple names and related non-geographic entities. Their work can be understood as an extension and a revision of OntoGazetteer, with a focus on the use of linked data to implement a gazetteer with worldwide coverage and able to keep information about every type of place, ranging from administrative divisions, populated places and intra-urban information, to geographic features, such as rivers and mountains.

This new gazetteer, built from the integration of many data sources on places, can be populated with related terms, alternative names, relationships of various kinds, and other related entities. In their implementation, they created an alternative schema for an enhanced gazetteer, which is more efficient for search and retrieval tasks. Figure 8.6 gives the structure of this system in which URI's refer to location in other gazetteers.

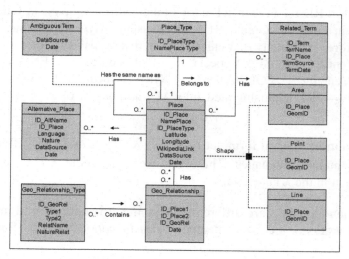

Figure 8.6. *Schema of the linked gazetteer from [MOU 16]*

In addition, the authors [MOU 16] claim that their system contains data on 13,074,366 places. More than 140 million place attributes and relationships were created. Furthermore, 4,477,739 non-place entities are available, having more than six million relationships with places.

8.4. Inference rules for matching geographic ontologies and gazetteers in different languages

In this section, we will consider matching GKS in different languages covering more or less the same territory. So, the conceptual framework will be given and will be followed by a few inference rules. However, before introducing them, let us explain language encoding and define homology relations for toponyms.

8.4.1. *Languages*

ISO 639 is a set of international standards that lists short codes for language names[8]. The following is a complete list of three-letter codes defined in part two (ISO 639-2) of the standard, including the corresponding two-letter (ISO 639-1) codes where they exist. In this chapter, we will use the three-letter codes as the prefix (ENG for English, FRE for French, ITA for Italian, SPA for Spanish, GER for German, GRE for Greek, RUS for Russian, ARA for Arabic, etc.). Therefore, for the city of Venice, we can distinguish various exonyms: ITA.Venezia, SPA.Venecia, FRE.Venise, ENG.Venice, GER.Venedig, POL.Wenecja, GRE.Βενετία, RUS.Венеция, ARA.البندقيـة (transliterated into Al Bundukiyya or Al Bondokia), etc.

In a more general form, let $\Lambda \equiv \{\lambda_1, \lambda_2, \lambda_3, ...\lambda_l: l \in N\}$ define the set of human languages. Therefore, we can denote $\lambda.Topo$ as any toponym in the λ language.

When alphabets are different, sometimes it is necessary to make a transliteration. Let us denote *Transliteration* as a function

---

8 http://www.iso.org/iso/home/standards/language_codes.htm.

transforming a text written in one alphabet into a text in a second alphabet according to rules.

Text2 = Transliteration $(\lambda, \quad Text1)$

8.4.2. Homology relation for toponyms

Indeed, regarding toponyms, equivalence relations can be created, for instance by writing "U.S.A." ≡ "United States of America". By extension, an equivalence class can be defined. However, for the translation of toponyms (Venice, Venezia, etc.), there is no systematic way to define them, and there is no authority to define them, except perhaps dictionaries. In other words, we are dealing with traditional translations agreed upon by many people. As a consequence, we can define a toponymic homology relation, such as ₪$_T$, so that we can write "Venezia" ₪$_T$ "Venice" or *Venezia* ₪ *Venice* to alleviate the notation. A homology class can be made by regrouping all of the agreed upon translations of a toponym. See Figure 8.7(a).

The case of the word "Monaco" is interesting. The Principality of Monaco (officially in French, Principauté de Monaco) is well known. However, in Italian, when the Italians speak about Monaco, there is an ambiguity, because there is another Monaco, namely Munich in Germany. To avoid the ambiguity, Italians speak about "Monaco di Baviera", so-called Bavarian Munich (Figure 8.7(b)).

Indeed, this relation is not an equivalence relation, because transitivity does not hold everywhere. Indeed, as previously explained, consider the Principality of Monaco (also called Múnegu in the Monégasque language, as given in Figure 8.7(b)). Therefore:

Monaco ₪ *Múnegu*

However, in the Italian language, the German city of Munich (München, in German) is also called Monaco. To distinguish, often the city is called Monaco di Baviera. Therefore, we can write:

Monaco ₪ *Munich*

It is obvious that "Munich" and "Múnegu" have nothing in common except their names in Italian. In the U.S., when speaking about a place named Washington, generally it is followed by the name of the state.

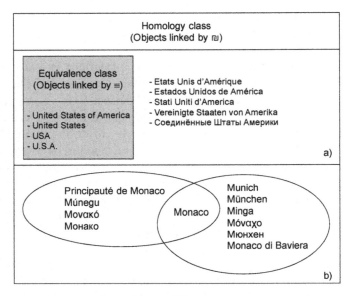

Figure 8.7. a) *Emphasizing the differences between equivalence and homology for toponyms; b) example of a toponym (Monaco) belonging to two homology classes*

8.4.3. *About exonyms and endonyms*

In addition, the various cities in the word Sevilla are also called Séville in French and أشبيليّة (transliterated into Ishbiliyya) in the Arabic language:

Sevilla ₪ *Séville*

Sevilla ₪ أشبيليّة

Regarding the multiplicity of languages, in some cases, it is important to get the endonym of a place. As previously mentioned, sometimes there are several possible endonyms: in this case, one will prefer the name as used by local people in a well-established language. For instance, in Canada, the French toponym "Québec" will be the endonym of the English toponym "Quebec." Therefore, let *Endonym* denote a function transforming any toponym into its corresponding endonym. For example:

$$Venezia = Endonym\ (Venice)$$

Back to the example of Granada, in order to disambiguate those homonyms, a solution is to use a hyperonym, for instance the name of the country as a topological prefix $\subset$ in which $A \subset B$ can be read as the combination of *Contains* (A,B) and *Covers* (A, B) or A is the hyperonym of B. Therefore, for distinguishing the several "*Granadas*" we can write ES $\subset$ Granada, US $\subset$ California $\subset$ Granada, US $\subset$ Kansas $\subset$ Granada, MX $\subset$ Granada, etc., in which MX stands for Mexico, ES for Spain. Let us call them located toponyms. As a consequence, if there is no ambiguity, we can define relationships in another mode:

$$RelationX\ (Topo_1,\ Topo_2)$$

As the previous solution is interesting to distinguish designated human settlements or administrative subdivisions, it cannot be used directly to disambiguate natural features, such as rivers, mountains, etc., which can spread over several cities, provinces, regions and even countries. Indeed, US $\subset$ Mississippi can relate to both the State and the river.

Let us define Earth as a toponym with its homologous "planet", "our planet", etc. Any located toponym can derive from Earth by a sort of inclusion path. For instance:

ENG.Earth $\subset$ "Pacific Ocean"

ENG. Earth $\subset$ America $\subset$ "North America" $\subset$ California $\subset$ Granada

ENG.Earth $\subset$ America $\subset$ "North America" $\subset$ "Lake Erie".

For the path description of Hawaii, there are two possibilities:

– country inclusion: ENG.Earth ⊂ "U.S.A." ⊂ Hawaii;

– location inclusion: ENG.Earth ⊂ "Pacific Ocean" ⊂ Hawaii.

8.4.4. Matching two geographic ontologies each in different languages

In this section, we will examine two geographic ontologies respectively designed in different languages (λ_1, and λ_2), for instance one in English and one in Spanish. Since concepts can be different or differently organized, how can we match them?

From a mathematical point of view, we have two graphs in which nodes correspond to concepts and edges to relations. Matching ontologies means that:

– types will be linked by homology relations;

– ontological relations will also be linked via homology relations.

8.4.5. Homologous geographic objects

Two geographic objects sharing homologous geometries, homologous toponyms and homologous types are said to be homologous (O_1 ⋈ O_2). In this case, they can be regrouped to form a single object having linguistic descriptions in two different languages. However, the newly-created object must have a unique geometric description. Several solutions can be given:

– adopt the more recent geographic description;

– adopt the more accurate;

– or create a mix of both.

8.5. Enriching geographic knowledge bases by rules

Based on the previous formalism, let us explain and write a few rules involving gazetteers and ontologies. Consider two geographic knowledge bases, each one developed with a different language. Suppose it is possible to transform them into the following structures:

$$GKB_1 \equiv (Terr_1, \ \lambda_1, \Omega_1, GO_1, \ \Gamma_1, \mathcal{R}_1)$$

$$GKB_2 \equiv (Terr_2, \ \lambda_2, \Omega_2, GO_2, \ \Gamma_2, \mathcal{R}_2)$$

With

$$GO_1 \equiv (GeoID_1, Geom_1, Topo_1, \Omega_1\text{-}Type)$$

$$GO_2 \equiv (GeoID_2, Geom_2, Topo_2, \Omega_2\text{-}Type)$$

in which languages (λ), ontologies (Ω), gazetteers (Γ), sets of geographic objects GO, and relationships ($\mathcal{R}$) are different ($\lambda_1 \neq \lambda_2$, $\Omega_1 \neq \Omega_2$, $\Gamma_1 \neq \Gamma_2$). However, in addition, the territories are supposed to have some parts in common (*Intersection*($Terr_1$, $Terr_2$) $\neq \varnothing$); otherwise, there is no way to compare or match them.

8.5.1. Inferring geometry

Suppose we have the description of two geographic objects each in one knowledge base, and suppose that one of the objects has an unknown geometry (noted *null*). If their toponyms and types are homologous, we can infer that those objects are homologous. In addition, we can transfer the geometry (Figure 8.8). Formally, we have:

$\forall O_1 \in GKB_1, \forall O_2 \in GKB_2, \forall \ \lambda_1, \lambda_2 \in \Lambda:$ $\lambda_1 \neq \lambda_2 \wedge (\lambda_1.Topo_1(O_1) \ ₪ \ \lambda_2.Topo_2(O_2))$ $\wedge \ (\Omega_1\text{-}Type(O_1) \ ₪ \ \Omega_2\text{-}Type(O_2))$ $\wedge \ (Geom_2(O_2) = null)$ $\Rightarrow$ $(O_1 \ ₪ \ O_2)$	Rule 8.1

In the case of ambiguities, for instance to decide among the possible rivers named Avon in the U.K., a solution can be to ask the user to help situate approximately within, for instance, a minimum bounding rectangle *MBR*. By doing so, the research space can be reduced until there is no ambiguity.

$\forall O_1 \in GKB_1,\ \forall O_2 \in GKB,\ \forall\ \lambda_1, \lambda_2 \in \Lambda:$ $\lambda_1 \neq \lambda_2 \wedge (\lambda_1.Topo_1(O_1) \bowtie O_2.\ \lambda_2.Topo_2(O_2))$ $\wedge\ (\Omega_1\text{-}Type(O_1) \bowtie \Omega_2\text{-}Type(O_1))$ $\wedge\ (Geom_2(O_2) = null)$ $\wedge\ (Contains\ (MBR,\ Geom_1(O_1)))$ $\Rightarrow$ $(O_1 \bowtie O_2)$	Rule 8.1b

Therefore it implies also $(O_1.Geom_1 \bowtie O_2.Geom_2)$. To reinforce the validity of the last relationship $(O_1.Geom_1 \bowtie O_2.Geom_2)$, since one of the starting value was originally unknown $(O_2.Geom_2 = null)$, the best solution is to copy the geometric value in order to give $(O_2.Geom_2 = O_1.Geom_1)$. Another possibility is to regroup both objects into a single one, so having two descriptions: in this case, the schema of the knowledge base must be modified accordingly.

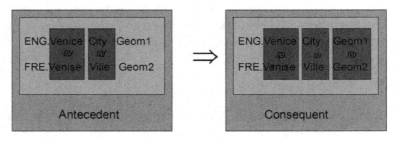

Figure 8.8. *Example illustrating Rule 8.1*

8.5.2. *From homologous geometry to homologous objects*

Consider now two objects with homologous geometries; we can infer that (see Figure 8.9): (*i*) their toponyms are homologous; (*ii*) their types are homologous; (*iii*) and so, the geographic objects are homologous. Formally, we can write:

$\forall O_1 \in GKB_1,\ \forall O_2 \in GKB_2:$ $Geom_1(O_1) \bowtie Geom_2(O_2)$ $\Rightarrow$ $\{(Topo_1(O_1) \bowtie Topo_2(O_2));$ $(\Omega_1\text{-}Type(O_1) \bowtie \Omega_2\text{-}Type_2(O_2));$ $(O_1 \bowtie O_2)\}$	Rule 8.2

As a consequence, by applying this rule, we generate correspondences in both gazetteers, and we provide a translation of two types in both ontologies.

Suppose you are in Finland, are Spanish-speaking and are facing the Lake Sääksjärvi: you will say in Spanish "Lago Sääksjärvi" or perhaps "Lago Saaksjarvi." In the case where a toponym is unknown, say $Topo_2$, without loss of generality, that is, $Topo_2 \notin \Gamma_2$, the Γ_1 missing toponym can be forced to be the endonym of the other: so:

$$Topo_2 = Endonym \ (Topo_1)$$

When the alphabets are different, some transliteration is needed into the λ_2 language, so that:

$$Topo_2 = Transliteration \ (\lambda_2, Endonym \ (Topo_1)).$$

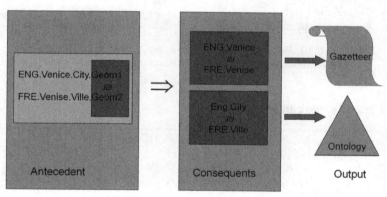

Figure 8.9. *Example illustrating the Rule 8.2*

8.5.3. *Inferring ontological relations*

Suppose now that, in addition, we have topological relationships in both gazetteers. Therefore, the knowledge bases are now constituted as follows, in which ρ_1 and ρ_2 stand for any type of ontological relationships:

$$GKB_1 \equiv \{O_1, \rho_1\} \text{ and } GKB_2 \equiv \{O_2, \rho_2\}$$

With $\rho_1 = R_1(GeoID_{11}, GeoID_{12})$ and $\rho_2 = (R_2GeoID_{21}, GeoID_{22})$.

If two couples of homologous objects have relationships between them, then their relations are homologous (see Figure 8.10 for an example in which Mediterranean and Adriatic Seas are linked by two different relationships).

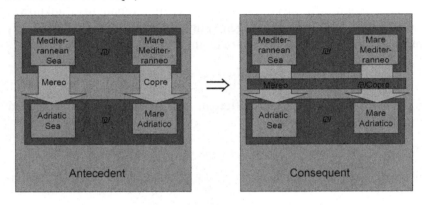

Figure 8.10. *Example of Rule 3 for inferring homology between two relations in which "Mereo" means a meronym relation*

Formally, we can write:

$\forall O_{11}, O_{12}, R_1 \in GKB_1, \forall O_{21}, O_{22}, R_2 \in GKB_2:$ $(O_{11} \bowtie O_{21}) \wedge (O_{12} \bowtie O_{22})$ $\wedge (R_1(O_{11}, O_{12})) \wedge (R_2(O_{21}, O_{22}))$ $\Rightarrow$ $(R_1 \bowtie R_2)$	Rule 8.3

An interesting case is when the relation name is unknown in one knowledge base, for instance, say R_2. In this case, there are several solutions:

– to confer the same name, but in this case, it is often not correct in the second language;

– or to ask an expert to propose a solution for the translation of the name of this relation.

Perhaps, some other rules can be written, so as to match geographic objects, their geometric shapes, their place names and types in different languages.

8.6. Conclusion

The goal of this chapter was twofold – first to present the problems of toponyms and the different ways to organize gazetteers, and second to establish the connections between geographic objects, ontologies and gazetteers in multilingual contexts.

Concerning gazetteers, until now, they are overall textual, but it could be of interest to include multimedia information such as (an example is provided in Figure 12.2):

– flags for countries and territories which have one;

– emblems, seals or logos;

– pictures of important landmarks,

– national hymns.

We have established some inference rules in order to match concepts between two geographic ontologies, each of them written with a different language. We have shown that gazetteers can be used in the foundation of this matching, not only for concepts, but also for relations between concepts. Several inference rules were described, but certainly some additional rules can be designed.

One of the main assumptions was that time was not involved. This can be a perspective to extend this framework in order to take temporal aspects into account. Among the difficulties, remember that toponyms and, more precisely, archeonyms can evolve, but the overall geographic descriptions of old features remain unknown or very difficult to estimate: what were exactly the coordinates of Roma as created by Romulus?

An additional remark could be of interest: it could be interesting to include pictures in a gazetteer to transform it into a multimedia gazetteer. Indeed, a lot of landmarks could be easily identified by pictures (see for instance the Eiffel Tower in Paris, or the Corcovado in Rio de Janeiro.

Another remark is that we have shown the following equivalence: two geographic objects are homologous when their geometry, ontological type and toponyms are homologous. Perhaps, in the future when dealing with time, this aspect must be integrated.

$$O_1 \bowtie O_2 \Leftrightarrow \left\{ \begin{array}{c} Geom\,(O_1) \bowtie Geom\,(O_2) \\ \Omega\text{-}Type\,(O_1) \bowtie \Omega\text{-}Type\,(O_2) \\ Topo\,(O_1) \bowtie Topo\,(O_2) \end{array} \right.$$

More, homology between relations was also defined ($R_1 \bowtie R_2$).

This chapter can also be seen as a first step towards the fusion of several geographic knowledge bases written in different languages.

Another perspective is to include non-spatial attributes, as they are very common in GIS. Due to this, the geographic knowledge base can be enriched by knowledge extracted from spatial data mining.

9

Geographic Knowledge Discovery and Data Mining

Where does knowledge come from? This is a very old question. One clue lies perhaps within databases under the assumption that knowledge is hidden in data as presented in Principle #1. Like miners trying to extract gold nuggets, computer scientists have developed theories and techniques to search databases in order to extract knowledge. Under the name of knowledge discovery in databases (KDD), these issues are now becoming very popular in a branch of IT called big data.

Historically speaking, one basis of knowledge discovery can dated to 1854 due to cholera in London. The Broad Street cholera outbreak was a severe outbreak of cholera that occurred near Broad Street in the Soho district of London, England. This outbreak is well known for the physician John Snow's study and his hypothesis that contaminated water, not air, spread cholera. John Snow made a map (Figure 9.1) with the location of pumps and deaths from cholera. In this map, one can easily see the correlations.

The scope of this chapter will be to briefly examine the various existing techniques for knowledge discovery from data mining and then to see how to apply them for geographic knowledge engineering.

However, algorithms will not be provided in detail and we ask the reader to refer, for instance, to [MIL 09] or [LI 15].

One aspect which is not dealt with in this book is geographic knowledge discovery from text mining. For the time being, only geographic information retrieval exists in texts or on the Internet (see [SAL 13], for instance, for details). Perhaps its successors will be geographic knowledge retrieval with an important aspect dealing with retrieval of geographic rules within texts.

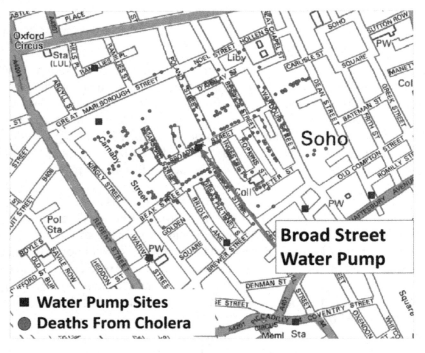

Figure 9.1. *Cholera London (from [SHE 16], reproduced with permission)*

9.1. Introduction to data mining

According to *Wikipedia*[1], data mining is the automatic or semi-automatic analysis of large quantities of data to extract previously unknown interesting patterns such as groups of data records (cluster analysis), unusual records (anomaly detection), and dependencies (association rule mining). This usually involves using database techniques such as indices. These patterns can then be seen as a kind of summary of the input data, and may be used in further analysis or, for example, in machine learning and predictive analytics. For example, the data mining step might identify multiple groups in the data, which can then be used to obtain more accurate prediction results by a decision support system or any system for territorial intelligence.

9.1.1. *KDD process*

KDD means Knowledge Discovery from Databases[2]. The scope of this process is to claim that knowledge is hidden in databases not only from expert's minds. But in order to run processes for extracting knowledge, it is necessary to have some preliminary steps. Indeed, there are algorithms and software products able to carry out these tasks, but they need to have data with certain formats. Some preliminary steps are hence necessary to run them as exemplified in Figure 9.2:

– selection: obtain data from various sources;

– preprocessing: cleanse data;

– transformation: convert to common format. Transform to new format;

– data mining: obtain desired results;

– interpretation/evaluation: present results to user in meaningful manner.

---

1 https://en.wikipedia.org/wiki/Data_mining.
2 https://en.wikipedia.org/wiki/Data_mining#Process.

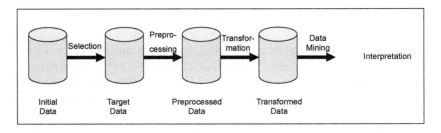

Figure 9.2. *The KDD process*

What kind of knowledge can be discovered by such a process? The idea is to discover regularities, frequent patterns, associations, correlations among sets of items or objects in transaction databases, relational databases and other information repositories. A frequent pattern can be defined as a set of items, sequence, etc. that occur frequently in a database. Among them, once discarding outliers, one can distinguish:

– association, correlation, causality;

– sequential patterns, temporal or cyclic association, partial periodicity, spatial and multimedia association;

– associative classification, cluster analysis, etc.

In the following sections, only associative rules will be detailed, first for common databases and then for spatial databases.

9.1.2. *Association rules in non-spatial databases*

Association rules are a class of regularities introduced by [AGR 93] that can be expressed by an implication of the form:

$P \Rightarrow Q$	
$[s, c]$	Rule 9.1

where P and Q are a set of items, such that $P \cap Q = \emptyset$, the parameter s, called support, estimates the probability $p(P \cup Q)$, and the parameter c, called confidence, estimates the probability $p(Q|P)$. We call an association rule $P \Rightarrow Q$ geographic, if $P \cup Q$ is a geographic pattern, that is, it expresses a relationship among geographic objects.

Suppose a New York City Tourist Office has a database concerning tourists and the landmarks they have visited. A fragment of 10 visitors is shown Table 9.1.

Name of Visitors	Lady Liberty	Ellis Island	MoMA	Empire State Building	Central Park
Ann	√	√			
Bill			√		
Bob	√	√		√	√
Fran	√	√		√	
Jill	√		√		√
Liz	√	√		√	√
Mary		√	√	√	√
Paul	√	√		√	
Pete	√	√	√	√	
Ted			√	√	√
Amount	7	7	5	7	5

Table 9.1. *Fragment of a tourist database*

Suppose we are interested by the link between Lady Liberty and Ellis Island. In this list, one can see that seven people visited Lady Liberty: so the support is 7/10 = 70%. Out of those persons, six visited also Ellis Island, so the confidence is 6/7 = 85%. So we can write:

Visit_Lady_Liberty $\Rightarrow$ *Visit_Ellis_Island* [Support = 70%, Confidence = 85%]	Rule 9.2

In addition, we can consider two columns. For instance consider Lady Liberty and Ellis Island versus The Empire State Building. In this case, the support of this couple is 6/10 = 60% and the confidence is 4/6 = 67%.

Visit_Lady_Liberty ∧Visit_Ellis_Island ⇒ Visit_Empire_State_Building [Support = 60%, Confidence = 67%]	Rule 9.3

Examining other databases, here are some examples of discovered associative rules. For instance, let us assume that in a database, the following association rule is discovered in a database of a computer shop:

Buy_Computer(X) ⇒ Buy_Antivirus_software(X) [Support = 20%, Confidence = 60%]	Rule 9.4

Which means that if somebody X buys a computer, there is a 60% confidence to buy also an antivirus software. Another example:

Age (X, "20..29") ∧ occupation (X, "student") ⇒ buys(X, "PC") [Support = 5%, Confidence = 70%]	Rule 9.5

This last associative rule means that considering a person who is a student and whose age is between 20 and 29 years then this person buys a PC with a support of 5% of all items sold in this company with a confidence of 70%.

Other additional measures of interestingness have been proposed by [GEN 06]. After this rapid introduction, let us treat spatial data mining[3].

---

3 Usually, the expression "spatial data mining" is used, but I prefer "geographic data mining".

9.2. Elements of spatial data mining

As previously explained, the methodology presented for extracting associative rule is not totally adequate for geographic knowledge. Indeed, the question addressed can be as follows:

– what are geographic objects which can be frequently associated?

– where will a phenomenon occur?

– which spatial events are predictable?

– how can a spatial event be predicted from other spatial events?

More precisely, here are some other more practical examples:

– where will an endangered bird nest?

– which areas are prone to fire, given maps of vegetation, draught, etc.?

– what should be recommended to a traveler to visit in a given location?

Domains	Example Features	Example of Co-location Patterns
Ecology	Species	(Nile crocodile, Egyptian plover)
Earth science	Climate and disturbance events	(Wild fire, hot, dry, lightning)
Economics	Industry types	(Suppliers, producers, consultants)
Epidemiology	Disease, types and environmental events	(West Nile disease, stagnant water sources, dead birds, mosquitoes)
Location-based Services	Service type requests	(Tow, police, ambulance)
Weather	Fronts, precipitation	(Cold front, warm front, snow fall)
Transportation	Delivery service tracks	(US Panel Service, UPS, newspaper delivery)

Table 9.2. *Examples of Co-location Patterns [SHE 06]*

Taking these issues into account, the goals of Spatial Data Mining can be listed as:

– identifying spatial patterns;

– identifying spatial objects that are potential generators of patterns;

– identifying information relevant for explaining the spatial pattern (and hiding irrelevant information);

– identifying co-location patterns, that is couples of geographic objects that occur frequently at the same location or within a small distance between them;

– removing outliers, that is geographic objects or attributes which are outside common ranges;

– designing good spatial clusters, that is regions with similar characteristics.

In order to do so, one has to define what a spatial pattern is. Beforehand, some phenomena cannot be identified as spatial patterns because they are random, casual or fortuitous. Consequently, here is a list of spatial patterns:

– a frequent arrangement, configuration, composition, regularity;

– a rule, law, method, design, description;

– a major direction, trend, prediction;

– a significant surface irregularity or unevenness.

9.2.1. Co-location patterns

Among them, the co-location patterns seem the more promising. [SHE 01] quotes a list of various domains in which co-location patterns were discovered (Table 9.2).

But some other kinds of geographic knowledge can be discovered as given in Table 9.3 according to Li and Wang [LI 05].

Knowledge	Description	Examples
Association rule	A logic association among different sets of spatial entities that associate one or more spatial objects with other spatial objects. Study the frequency of items occurring together in transactional databases.	*Rain (x, pour) => Landslide (x, happen)*, support is 76%, and confidence is 51%.
Characteristics rule	A common character of a kind of spatial entity, or several kinds of spatial entities. A kind of tested knowledge for summarizing similar features of objects in a target class.	Characterize similar ground objects in a large set of remote sensing images.
Discriminate rule	A special rule that tells one spatial entity from other spatial entity. Different spatial characteristics rules. Comparison of general features of objects between a target class and a contrasting class.	Compare land price in urban boundary and land price in urban center.
Clustering rule	A segmentation rule that groups a set of objects together by virtue of their similarity or proximity to each other in the unknown contexts what groups and how many groups will be clustered. Organize data in unsupervised clusters based on attribute values.	Group crime locations to find distribution patterns.
Classification rule	A rule that defines whether a spatial entity belongs to a particular class or set in the known contexts what classes and how many classes will be classified. Organize data in given/supervised classes based on attribute values.	Classify remotely-sensed images based on spectrum and GIS data.
Serial rules	A spatiotemporal constrained rule that relates spatial entities in time continuously, or the function dependency among the parameters. Analyze the trends, deviations, regression, sequential pattern, and similar sequences.	In summer, landslide disaster often happens. Land price is the function of influential factors and time.
Predictive rule	An inner trend that forecasts future values of some spatial variables when the temporal or spatial center is moved to another one. Predict some unknown or missing attribute values based on other seasonal or periodical information.	Forecast the movement trend of landslide based on available monitoring data.
Exceptions	Outliers that are isolated from common rules or derivate from other data observations very much	A monitoring point with much bigger movement.

Table 9.3. *Main spatial knowledge to be discovered, according to [LI 05]*

9.2.2. Association rules extracted from spatial data mining

As previously enumerated, the major families of data mining applications allow to determine the outliers and relationships of co-locations, spatial correlations and clustering. In Figure 9.3, [YOO 14] gives the main patterns which can be discovered by spatial data mining.

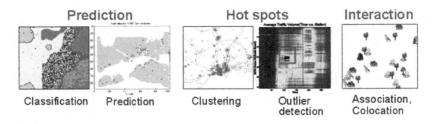

Figure 9.3. *Spatial pattern families versus techniques. From [YOO 14]. Reproduced with permission. For a color version of the figure, see www.iste.co.uk/laurini/geographic.zip*

Here come a few examples:

a) This first example is taken from [APP 03]. For instance, a user may want to discover the spatial associations of towns in British Columbia with roads, waters or boundaries having some specified support and confidence.

Finally, the following spatial association rule has been discovered (in which *DB* means the concerned database):

$\forall X \in DB,\ \exists Y \in DB,\ \textit{is-a}\ (X,\ town)$ $\Rightarrow$ *close-to* $(X,\ Y) \wedge$ *is-a*$(Y,\ water)$ [Confidence 80%]	Rule 9.6

This rule states that 80% of the selected towns are close to water, that is the rule characterizes towns in British Columbia as generally being close to some lake, river etc.

b) Co-locations rivers and roads.

This example is taken from [SHE 06] from a dataset in Korea in which the authors look for links between:

– road: river/stream;

– crop land/rice fields: ends of roads/cart roads;

– obstacles, dams and islands: river/streams;

– embankment obstacles and river/stream: clayey soils;

– rice, cropland, evergreen trees and deciduous trees: river/stream;

– rice: clayey soil, wet soil and terraced fields;

– crooked roads: steep slope.

Figure 9.4 gives the result of this study concerning rivers and roads.

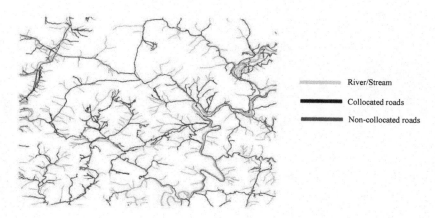

River/Stream
Collocated roads
Non-collocated roads

Figure 9.4. *Co-location between Roads/Rivers-Streams. From [SHE 06]. Reproduced with permission. For a color version of the figure, see www.iste.co.uk/laurini/geographic.zip*

9.3. Conclusion

Starting from the idea that knowledge is hidden in databases, spatial data mining is now a very important domain of research.

Among the extracted knowledge bunches, co-location rules seem to be emerging as the more interesting ones.

Do not forget that other sources of geographic knowledge can come from the analysis of books and websites under the expression of text mining, or more specifically, geographic text mining.

Geographic Applicative Rules

According to [GRA 06] and [MOR 08], rules (business rules) should be considered as first-class citizens in computer science. In enterprises, the "craft" of expert know-how is capitalized in an information system in the form of "business" rules. These rules can then be explained and implemented in applications such as business intelligence in software architectures integrated type ERP (Enterprise Resource Planning) or not. For example, the SAP-integrated software package is based on a declarative formalism for the description of the job tasks with lists of rules, such as "automobile insurance does not cover drivers who have been recognized guilty of driving while intoxicated over the past two years", or "when monthly invoices are sent, the supplementary documents that match the profile of the client should be included". On the other hand, the explanation and the formalization of business rules is still a hot topic as a new standard from the OMG was issued in September 2015 entitled Decision Model and Notation. The objective is to provide a language of formalization of business rules including those exploited in the decision-making process (OMG - DMN 2015)[1].

Thus, the goal of this chapter[2] will be to review the rules of geoprocessing in order to extract particular semantics and to lay the foundations of a machine-processable model. Voluntarily, I will tackle

---

1 http://www.omg.org/spec/DMN/.
2 Preliminary versions were published in [LAU 15c] and [LAU 16b].

neither temporal aspects nor 3D aspects, although sometimes it is necessary to use these dimensions in certain rules.

The areas of application fall within geoprocessing. Specifically, we are interested in the rules concerning not only geographic objects, but all objects whose knowledge of the location is important.

In the previous chapters, several rules were presented, but they are not applicative rules. Indeed, they are more basic rules which constitute a sort of layer which applicative rules can use.

This chapter is thus built: after having presented the importance of rules in information technologies, an introductory example and detailed computer modeling of many rules will be examined. Finally a sketch of the model will be given and some different perspectives will be explored. And to conclude this analysis, a definition of applicative geographic rules will be presented.

10.1. About rules in information technology

A rule is a basic element of a strategy to build reasoning. In contrast to algorithms, they are expressed declaratively. Among business rules, Dietz [DIE 08] distinguishes between three categories:

– *rejectors*, typically those related to quality control, that allow a rejection (rejection rules);

– *producers*, such as those determining new values (e.g. VAT calculation); they can be considered as rules of production of information;

– and *projectors* such as those related to the replenishment of stocks.

In our case, a rule is not necessarily a legal regulation, but simply an inference (implication) between elements or phenomena whose origin can also be physical, statistical or best practice type, or from data mining.

Few studies have been conducted on automatic reasoning in geoprocessing. Nevertheless we could consider the works presented in the book edited by [KIM 89] and the special issue of the CEUS journal, including the editorial of [BAT 91], which were very typical of the time. But these works are old and the knowledge engineering framework has drastically changed. More recently, one could mention the paper written by [JAI 11], but it does not address the reasoning part based on declarative rules.

Let us also consider the paper of [SAL 15] which operates a GIS on minerals data to discover characterization rules that can facilitate data mining and spatial analysis. In the same spirit, [BIA 16] present methods for extracting the characteristics of a mountainous terrain (peaks, passes, edges, etc.) from the raw digital terrain models, which could easily be described in the form of rules.

However in geoprocessing, beyond administrative rules, other statements may lead to geographic applicative rules. In fact, let us look at some of them:

– in the United Kingdom, they drive on the left;

– in Canada, the majority of the population lives along the border with the United States;

– each capital city has an international airport nearby;

– between two capital cities, in general, there are direct flights;

– in the Northern Hemisphere, the further North you go, the colder the weather (but locally this is not always true);

– the higher you climb a mountain, the lower the temperature;

– heavy rain upstream, downstream flooding;

– mosques are oriented towards Mecca;

– if a zone is a swamp, it is necessary to prohibit construction;

– if there is unemployment, the creation of companies or industrial areas must be encouraged;

– if a plot is adjacent to an airport, it is necessary to limit the height of buildings;

– it is forbidden to open a new tobacco shop within 500 m of an already existing one;

– in air-polluted and windy areas, wind spreads air pollutants;

– when you want to install a metro-line under a street, please move other underground networks to another place;

– a good practice in Mexico is to take a bus to go from Puebla to Oaxaca City.

Another example is derived from the use of electrical plugs in different countries. This table could be considered as a visual decision table or a decision map (Figure 10.1).

Among the rules, there are several categories. Take two examples "if it is raining, I get wet" and "if it is raining, I take an umbrella". In the first case, it is a rule of physical type whose consequence is systematic provided that I am outside. For the second, it reveals a good practice that I am not obliged to follow. Therefore, we can see that if the premises are identical, the status of the conclusions can be totally different.

As previously mentioned in section 2.4.5, [BOL 10] suggested several XML extensions to model rules. The simplest of these is as follows:

```
<Implies>
<if>
<..>
</if>
<then>
<..>
</then>
</Implies>
```

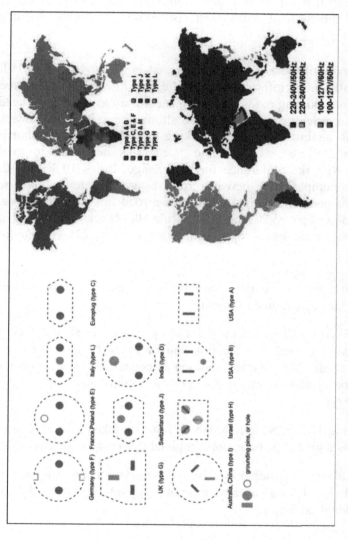

Figure 10.1. *Example of a decision map for electrical plugs and the countries where they are standardized. Source: ttp://wikitravel.org/en/Electrical_ systems. For a color version of the figure, see www.iste.co.uk/laurini/geographic.zip*

10.2. Introductory example regarding street naming

In [VAR 16], an example of rule encoding is given concerning road naming in Australia in order to automate the process. Rules are defined in the form of ontological vocabularies using the language SWRL[3].

Only some of these rules are presented in Table 10.1. Rule 10.1 automatically infers information with the help of a road link between proposed and existing roads; this rule is necessary as every road needs to link with at least one other road to allow access. Rule 10.2 checks road length against road type; checking the road length for shortest road types ("Place", "Close" and "Lane") is necessary to avoid confusion with the preference for road usage. Rules 10.3 and 10.4 check the compatibility between road usage and road links; for example an open-ended road must have a road link at both the start and end point of the road. And finally Rule 10.5 checks whether or not the proposed road has a wide panoramic view across surrounding areas.

However several remarks can be made because this is logic reasoning, not geographic reasoning:

– Rule 10.1 tests whether the new proposed road is linked to another existing road; but the link can be made by a set of newly proposed roads. The adapted solution is concerned by the order of presentation of new streets. A very general solution must be based on graph theory;

– in Rule 10.2, road length, which must be taken into account, is provided as a given attribute, not computed from road coordinates;

– in Rule 10.5, panoramic view is also given as an attribute, not calculated by taking terrain morphology into account by 3D computational geometry.

---

3 https://www.w3.org/Submission/SWRL/.

Rule 10.1: Relate a road link with existing road directly through another proposed road	Road(?R1), Road(?Old), has RoadLink(?R1, ? Old), status(?R1, "New"), status(?Old, "Existing"), notEqual(?R1, ?R2) -> isAllowed(?R1, true)
Rule 10.2: Check the road length against road types	RType(?T1), Road(?R1), Road_Type(?R1, ?T1), hasLength(?R1, ?200), SameAs (?T1, ?$Close$) -> isAllowed(?R1, true)
Rule 10.3: Check the road access against road type	Road(?R1), hasRoadUse(?R1, "Openended"), Road(?Old1), Road(?Old2), hasRoadLinkS(?R1, ?Old1), hasRoadLinkE(?R1, ?Old2), status(?R1, "New"), status(?Old1, "Existing"), status(?Old2, "Existing"), notEqual(?R1, ?Old1), -> isAllowed(?R1, true)
Rule 10.4: Check the road usage against road link.	Road(?R1), hasRoadUse(?R1, "cul-de-sac"), Road(?Old1), Road(?Old2), hasRoadLinkS(?R1, ?Old1), hasRoadLinkE(?R1, ?Old2), status(?R1, "New"), status(?Old1, "Existing"), status(?Old2, "Existing"), notEqual(?R1, ?Old1), -> isAllowed(?R1, false)
Rule 10.5: Check the roadway with view	RType(?T1), Road(?R1), Road_Type(?R1, ?T1), SameAs(?T1, ?$Vista$) -> isAllowed(?R1, true)

Table 10.1. *Example of rules taken from [VAR 16]*

10.3. Geographic knowledge and reasoning

The objective of this section is to define the threshold between the bases of geographic knowledge and automatic reasoning. After some generalities, we will quickly review the structuring of the knowledge bases and computer modeling of rules.

10.3.1. *General information*

The purpose of a declarative rule-based model is to allow automatic reasoning. In contrast to expert systems of the past that were simply using logic, in our case, it will be very different. Here are some examples (section 3.2):

– defining the location of a new airport, a new hospital, a new stadium, social housing, etc.;

– checking the compliance of a building vis-à-vis building regulations;

– determining the best way to go from A to B;

– organizing a policy about urban green spaces;

– determining transit policy;

– limiting crime in a city;

– planning a tour through X and Y;

– planning garbage collection circuits;

– determining the most polluted, noisiest, places;

– organizing the evacuation of people during a volcanic eruption;

– organizing relief after an earthquake;

– etc.

From these examples, we find that geographic reasoning must invoke other mathematical disciplines such as:

– spatial reasoning through the integration of topology and computational geometry;

– graph theory in some cases like searching for routes or circuits;

– spatial analysis;

– simulation techniques;

– fuzzy logic and reasoning;

– multi-criteria decision theory;

– and operations research.

These elements can be integrated in the form of procedures which will be invoked on-demand.

10.3.2. *Geographic knowledge bases*

Once having selected language and territory, remember (Figure 3.4) that a geographic knowledge base may include an ontology, a gazetteer, geographic objects, the relationship between these objects, rules and mathematical models. Figure 10.2 (copy of Figure 3.7) shows such a base linked to a geographic inference engine that will allow reasoning.

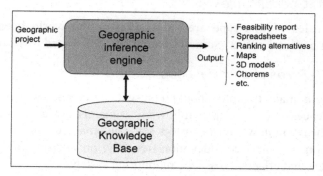

Figure 10.2. *Linking a geographic inference engine to a geographic knowledge base*

In input, there will be a territorial project to study so as to determine, through the geographic inference engine, the consequences in output. These consequences will be grouped by feasibility which may be in the form of maps, chorems (see section 11.3), diagrams, texts, etc.

To conclude this section, all the components of a geographic knowledge base can be used not only to write a rule but also in their activation context.

10.4. Study of the semantics of the geographic rules

Now let us examine certain rules in different areas. In the previous chapter, many generic rules were presented, but the scope of this section is to study applicative rules. First, we will discuss the rules across the globe (often of a physical nature), then those only valid in certain places (often of an administrative nature) and those coming from data mining.

10.4.1. *Global rules*

In this category, one may include geodetics and rules of physical geography.

10.4.1.1. *Geodetic rules*

The rules of this type are valid all over the world because we consider cardinal points. Some of them were developed in section 5.3.

10.4.1.2. *Rules of physical geography*

In this domain, the rules should represent natural phenomena and their consequences. For example following tsunamis, volcanic eruptions, storms, heavy rain, we must consider some automatic consequences. But, in addition, we must consider more recently, prevention, protection or mitigation, and effective real-time monitoring systems.

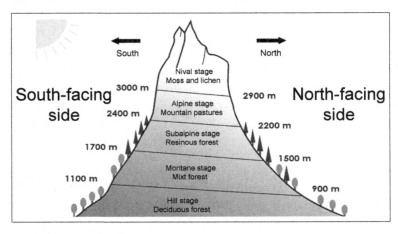

Figure 10.3. *Example of vegetation layers in the Alps. Source: translated from https://www.jardinalpindulautaret.fr/jardin/cadre-naturel-exceptionnel/letagement-vegetation-en-montagne*

But due to the local topography, some rules may be invalidated. In the northern hemisphere, the further north you go, the colder it becomes. On a certain scale, this type of rule is valid, but there are places where this reasoning is no longer valid. Therefore, one must distinguish between local and global rules. Here the local rule supersedes the global rule as, for example, when one is dealing with microclimates.

The rules of the spatial distribution of flora and fauna, hydrology, etc. from climatology and meteorology fall into this category. Figure 10.3 gives an example of vegetation layers in the Alps.

Suppose that we are on the southern slopes (sunside): two options to design a rule are rapidly written:

| IF Lichen THEN Altitude > 3000; | Rule 10.6 |
| IF Altitude > 3000 THEN Lichen. | Rule 10.7 |

In the first case (Rule 10.6), it would be a study linking a type of vegetation to elevation, while the second (Rule 10.7) shows since we are at some level from the sea, what the types of flora that we can meet are.

Now consider the case of mathematical models and assume that we have $A = M (B, C, D)$. This formula can be easily transformed into a rule in the following manner: one must:

– write a procedure or a function representing M, which will be encapsulated;

– search for or determine B, C and D in the antecedent part;

– then run M in the consequent part in order to determine A.

Another variant would be that A may intervene into a condition; therefore M will be invoked in the antecedent part.

10.4.1.3. *Rules in urban civil engineering*

In this domain, many rules exist. Among others, let us mention (for designing and maintenance):

– rules for water catchment and supply;

– rules for waste water cleaning;

– rules for gas and electricity distribution, energy supply and district heating;

– rules for organizing public transportation;

– rules for organizing sensor networks;

– rules for waste collection and management;

– and for any kinds of public utilities.

10.4.1.4. Rules dealing with public services

Similarly, there are administrative rules such as those for schools and various levels of education, for hospitals, for social and medical-social centers and for all other services proposed to citizens. Often, they are based on the number of people living in a zone.

Remember that the quality of those services and also of public utilities is a key-element in a smart city.

10.4.2. Local rules

Here, we will discuss only the rules applied on a restricted territory, namely the administrative rules and those relating to specific spaces.

10.4.2.1. Rules coming from laws

Each country has its own rules, not only from an administrative point of view, but also from location. For instance, when analyzing road traffic from aerial photos, it is important to know that in the United Kingdom the cars drive on the left. In addition, concepts such as language and currency can impact geographic rules.

In addition, typically in each country there is a Constitution and many laws regulating the geographic aspects related to urban and environmental planning. Take the example of tobacconists in France for whom it is forbidden to open a new tobacco shop less than 500 meters from an existing one. Figure 10.4 shows a visual example of this rule. To establish where it is possible to create a new tobacco shop (Z), several geometric operations such as the determination of buffer zones and geometric difference have to be implemented in an encapsulated way.

The rule for determining Z can be written as follows:

$$\forall\ F_i \in GO,\ \exists\ Z \in Terr,$$ $$G\text{-}Type(F_i) = Point,\ G\text{-}Type(Z) = Area,$$ $$\Omega\text{-}Type(F_i) = \text{“Tobacconist”},$$ $$Geom\ (F_i\ ) \in Terr$$ $$\Rightarrow$$ $$Geom(Z) = Terr - Union(Buffer(F_i,\ 500))$$	Rule 10.8

If somebody has a project to open a tobacco shop in Z, it will be approved regarding this rule. Maybe some additional rules may apply, for instance concerning the size and the precise location of the premises.

Moreover, one of the peculiarities of the rules of administrative origin is the existence of sanctions whenever they are not met. But there may be cases of exemptions, which should be taken into account in one way or another.

On the other hand, although there are international standards for the Highway Code, each country has its own peculiarities concerning priority at intersections, roundabouts, etc.

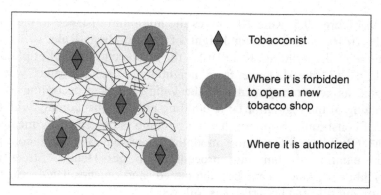

Figure 10.4. *Example of administrative rule: "it is forbidden to open a new tobacco shop within 500 m of another existing one"*

10.4.2.2. *Urban planning rules*

Generally, in each country, there are also laws that govern urban planning. Let us take a small example taken from building licenses as presented in Figure 10.5 where you can see a building that must follow certain rules.

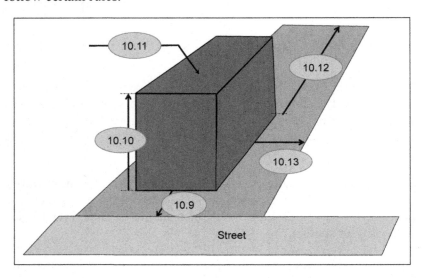

Figure 10.5. *Example of building-related planning rules*

In Figure 10.5, Rule 10.9 gives the minimum distance to the road; Rule 10.10, the maximum height of the building; Rule 10.11, the volume of the building; Rule 10.12, the distance to the end of plot and Rule 10.13, the distance from neighbors. Here, the rules will be treated as constraints, that is to be valid (accepted according to the meaning of the regulation), the building project must comply with this set of constraints. Suppose that the building is described by means of BIM (Building information modeling)[4], which is a very common representation of building, procedures to get Height, Street and Neighbor distance can be automated and encapsulated. So the following rule can be written as follows:

---

4 http://www.iu.edu/~vpcpf/consultant-contractor/standards/bim-standards.shtml.

$\forall\ B \in PROJECT, \exists\ P \in GO$ $\Omega\text{-}Type(B) = $ "Building", $\Omega\text{-}Type(P) = $ "Parcels", $Contains\ (Geom(P),\ Geom(B))$: $Height(B) < 10$ $\wedge\ Street\_distance(B, P) > 3$ $\wedge\ Neighbor\_distance(B, P) > 3$ $\Rightarrow$ $UP\text{-}Allowed\ (B, P)$	Rule 10.9–10.13

Among the urban planning rules, there may be good practices such as "burying the engineering networks (electricity, telephone)" or even "before creating an underground metro line, we must move sewerage networks".

10.4.2.3. Local socio-economic rules

The majority of countries have developed rules for the organization of the economy and companies. These rules have a significant impact on the use of the land. Figure 10.6 provides an illustration of the rule "along the edges of sea, the greater the price, the greater the distance from the sea, the lower the prices of homes are".

A particular case regards rules related to the flow of people or goods. Here two places or two families of places may intervene, either as origin or destination.

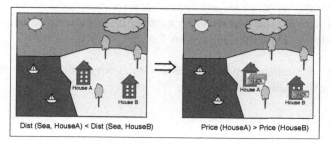

Figure 10.6. *Example of economic rule: "along the edges of sea, the greater the distance from the sea, the lower are prices of homes"*

10.4.2.4. *Good or best practices*

Among the rules of good practice, there are those related to the description of itineraries. In most countries where the landscape is densely populated, description is often done according to the directions, cities and villages to cross. In other countries where the landscape is lightly populated, the description is made by taking into account numbers and route directions, North, South, East and West. In deserts and seas, it was common to use the positions of stars as landmarks. These good practices had been used for centuries; now there are other ways to do so. However this type of knowledge strays from the topic of our book because it uses information outside the terrestrial globe (extra-terrestrial knowledge).

Good practices include techniques of numbering houses in cities, which may vary depending on the country. Alternatively, along a highway, the creation of an additional interchange may be the basis of a new community development or a new industrial zone.

10.4.2.5. *Cross-border rules*

Consider rules such as:

– in Geneva, Switzerland, there are suburbs in France; since wages are higher in Switzerland, many people live in France and work in Geneva;

– at the Libyan border of Tunisia, gasoline is less expensive.

From a more general point of view, they have the characteristics to link geographic objects, attributes, placenames located in both the internal jurisdiction and its vicinity: they constitute some neighboring knowledge as explained in section 3.4.

As a consequence, when necessary the geographic inference engine must consider cross-border rules.

10.4.3. *Low level or generic rules*

In this category, one can find the rules related to the acquisition and visualization of data. Among these, one can discuss three types of rule:

– those related to data quality control as presented in section 4.7.1;

– those related to the transformation of geometric objects (section 4.3);

– those related to topological relations (section 5.6) according to scales.

10.4.4. *Rules and plurality of places*

Four cases are to be analyzed:

a) From the previous examples, we can see that most of the rules refer a unique place. But a rule such as "in England, we drive on the left" also applies in other countries (Figure 10.7);

b) However the rules related to the flow of people, goods and animals are characterized by two places, the first so-called origin and the other, destination. Consider for example those governing the movements of migratory birds. Accordingly the grammar of the rules must allow this scenario by considering three cases:

– bipolar flows (the most common),

– diverging flows where only the source is known, for example emigration,

– and the converging flow where only the destination is known, as for immigration.

c) A third case is that of the rules corresponding to clusters of areas according to certain criteria, such as, for example, research of homogeneous areas in geomarketing or the definition of electoral boundaries. It is noteworthy that identifiers of these newly-created areas need to be assigned;

d) And finally another example is that of the importation of good practices from one place to another, for instance a new urban experiment.

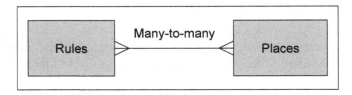

Figure 10.7. *Rules and places*

10.4.5. *Rules and logics of stakeholders*

One of the difficulties is the fact that among the urban actors, some have different "logics". With regard to industry creation, an environmentalist or an industrialist may have different ideas on the possible implications of this or that choice. Similarly, some groups may have different priorities (see Rule 10.14): facing a big empty space, athletes imagine a stadium, pupils' parents a school and a land developer a building, etc.

From the formal point of view, these aspects will occur in multi-actor and multi-criteria decision support systems.

$\forall P \in GO, \forall B \in PROJECT,$ $\Omega\text{-}Type(P) = \text{"Parcels"},$ $P.Landuse = \text{"Vacant"}:$ $Area(P) > 1000$	Rule 10.14
For an environmentalist	$\Rightarrow \Omega\text{-}Type(B) = \text{"Recreational\_Park"}$
For a sportsman	$\Rightarrow \Omega\text{-}Type(B) = \text{"Stadium"}$
For parents of pupils	$\Rightarrow \Omega\text{-}Type(B) = \text{"School"}$
For The Chamber of Commerce	$\Rightarrow \Omega\text{-}Type(B) = \text{"Start\_up\_facility"}$
For a land developer	$\Rightarrow \Omega\text{-}Type(B) = \text{"Residence"}$
For young parents	$\Rightarrow \Omega\text{-}Type(B) = \text{"Kindergarten"}$

10.4.6. *Handling exceptions*

The reality is always more complex than our models. Although there are no exceptions in physical rules, it is different for rules made by humans for several reasons. For instance in urban planning:

– due to corruption, laws or decrees are not followed in a lot of places;

– sometimes derogations or exemptions are granted;

– since rules change over the years, some buildings constructed according to laws in the past, do not follow today's laws;

– etc.

From a practical view point, in any geographic inference engine those aspects must be integrated, for instance by stating that in a certain place, a rule cannot be applied or can be superseded.

10.5. Toward applicative geographic rules modeling

Now that various examples have been analyzed, it is possible to extract elements of modeling. Firstly, general considerations will be given, and then a computer model will be proposed.

10.5.1. *General considerations*

As we saw earlier, new concepts have emerged and it is necessary to clarify what is meant by superseded rules, metarules, jurisdiction, etc. Finally two tables will make it possible to synthesize the characteristics of rules and our level of knowledge as well as their formalization.

10.5.1.1. *Superseded rules*

Indeed, certain rules can be superseded locally. In other words, it will be necessary to take this aspect into account not only in the design of the rules, but also in the inference mechanism.

10.5.1.2. *Metarules*

A metarule is a rule which offers a regulatory framework against which other rules must comply. For example, all local urban plans must be in compliance with higher level regulations, which thus appear as metarules. In other words, a metarule defines a set of rules that will be valid only when you refer to this metarule. In addition, it can define new concepts, new legal mechanisms, or even new decision-making bodies; a metarule can therefore enrich an ontology with this new terminology; this point, although current, is extremely complex and will not be tackled.

10.5.1.3. *Jurisdiction*

One can call jurisdiction, the territory of application of a rule, a metarule and even the entire knowledge base. Therefore, the gazetteer will only deal with place names within this jurisdiction or through it (rivers, roads, etc.). In some cases, it would be advisable to include close external information such as the names of neighboring places.

10.5.1.4. *IF-THEN-Zone rules*

As seen previously, the geographic rules commonly involve geographic objects (e.g. buildings in flood zone) and also geographic objects; see for instance vegetation (Figure 10.4) or habitats of animals.

But in addition, geographic objects may be deducted from rules. Let us call them IF-THEN-Zone rules. Take the example of maritime laws that distinguish territorial waters and international waters. The principle is based on the distance of 200 nautical miles[5] except for particular cases. And the rest is set as international waters. If we call *SEAS*, sea spaces, *OCEANS* ocean spaces, and *INTER_WATERS* international waters, one has the following geometric definition which will involve a geometric type out-buffer operation outside:

---

5 Remember that a nautical mile corresponds to 1,852 m. So 200 nautical miles correspond to 370,400 m.

$\forall\ c_i \in GO$, $\Omega\text{-}Type(c_i) = $ "Country", $G\text{-}Type(c_i) = Area$, $\exists\ CONTINENTS \in Earth$, $G\text{-}Type(CONTINENTS) = Area$, $\exists\ INTER\_WATERS \in Earth$, $G\text{-}Type(INTER\_WATERS) = Area$ $\Rightarrow$ $\{Geom\ (CONTINENTS) = Union\ (Geom(c_i))$; $Geom\ (INTER\_WATERS) = Union\ (Geom\ (SEAS)$, $Geom\ (OCEANS))\text{-}Buffer\ (Geom\ (CONTINENTS),\ 370400)))\}$	Rule 10.15

Similarly, it is possible to define inland waterways: suppose that they should be at least 5 m wide and 3 m deep, we could write naively:

$\forall r_i \in GO$, $\forall\ c_i \in GO$, $\Omega\text{-}Type(r_i) = $ "River", $G\text{-}Type(r_i) = Network\_of(r_i,\ confluence\_nodes)$, $\Omega\text{-}Type(c_i) = $ "Canal", $G\text{-}Type(c_i) = Network\_of(c_i,\ confluence\_nodes)$, $\exists\ Waterways \in Earth$, $G\text{-}Type(Waterways) = Network\_of(r_i \cup c_i,\ confluence\_nodes)$: $(Waterways.width > 5) \wedge (Waterways.depth > 3)$ $\Rightarrow$ $Geom(WATERWAYS) = Union\ (Geom\ (r_i),\ Geom\ (c_i))$	Rule 10.16

The big problem is that the variables *depth* and *width* are not known explicitly, but from a 3D geometric reasoning based on the morphology of the waterway. Figure 10.8 illustrates this problem. Furthermore, if the river has narrow meanders, it must ensure that long barges can pass; this aspect must also be integrated into Rule 10.15.

Ultimately, we deal with producing rules according to Dietz' terminology [DIE 08]. Generalizing the previous examples, we need to integrate the items deducted geometrically from geographic rules, namely new objects, new types of objects, new attribute values, or even new relationships between two objects and new spaces that can intervene for example as jurisdiction.

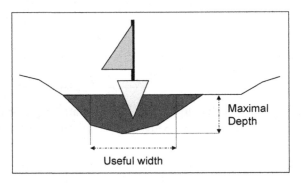

Figure 10.8. *Towards a definition of inland waterways*

Thus, once the rule for the determination of these objects is known, we can create a new class from this geographic rule. In other words, the consequent part of the rule will enrich the ontology by the creation of a new class derived from geometric reasoning, and can sometimes even enrich the gazetteer.

10.5.1.5. *Summary*

All these cases can be synthesized in two tables. Table 10.2[6] provides guidance on the nature of the rules and their characterizations: indeed, according to their origin, they have different features. The last column expresses the fact that the rule may be locally superseded by another.

The second table (Table 10.3) finally expresses the level of awareness that we have about the rules. At the level of formalization,

6 I am indebted to Sylvie Servigne and Franck Favetta for the construction of Tables 10.2 and 10.3.

rules of physical geography can be encapsulated into programs from a procedural way and, we need a mechanism to integrate them into a declarative model. With regard to the laws, they are known at a time *t*, but may change over time in the form of statutory instruments for which the translation into declarative forms may be difficult or too simplistic.

Characterization Origin of the rules		Emergence of rules Manual or automatic	Source Natural or artificial	Scope With or without local exceptions
Physical	Geodesy	Manual	Natural	Without
	Physical geography	Manual	Natural	With
Social	Juridical (including land use planning)	Manual	Artificial	With
	Socio-economy	Manual	Natural/Artificial	With
	Good practices	Manual	Artificial	With
Computing	From Data Mining	Automatic	Natural/Artificial	With
	Linked to quality control	Manual	Natural/Artificial	With
	Linked to visualization	Manual	Natural	Without

Table 10.2. *Characterization of rules*

In addition to the previous features, we need to clarify the nature of the implication ($\Rightarrow$). In fact, it has several meanings:

– in the case of physical phenomena, it corresponds to physical laws or causal chains (cause and effect); the implication is therefore automatic, sometimes with delay;

– however, if the physical law is only known empirically, the $a = f$ (b, c, d, etc.) type formula, where a is the value of an attribute of a geographic object, then the rule will affect the value calculated within a margin of error;

– legislative type rules are human laws; generally if the law is not obeyed, sanctions may appear, thus involving an ELSE clause in the computer rule;

– in association rules (sometimes referred to as frequent association) from data mining, the semantics of the $\Rightarrow$ sign should be modulated according to the value of confidence related to this association rule;

– concerning good practice rules, the implication will be judged in any manner desired;

– finally, if the rule involves fuzzy objects, the semantics of the $\Rightarrow$ sign will be modulated according to the values of fuzzy membership degrees.

Now that the main elements of geographic semantics have been identified, it is possible to propose a preliminary model.

Rules Origin of rules		A priori known Yes/No	Formally written Yes/No	Automated Yes/No
Physical	Geodesy	Yes	Partially	Partially
	Physical geography	No	Partially	Partially
Social	Juridical (including landuse planning)	Yes	Yes	No
	Socio-economy	No	Partially	No
	Good practices	No	Partially	No
Computing	From Data Mining	No	Yes	Yes
	Linked to quality control	No	Partially	Partially
	Linked to visualization	No	Yes	Yes

Table 10.3. *Levels of awareness and automation of rules*

10.5.2. *Outline of a model*

From the analyzed examples, first there is a many-to-many relationship between the rules and the names of places, and another between the rules and the types of geographic objects.

In addition, some territories (such as countries) have the ability to emit metarules which will apply on inner places and will be a normative framework for located rules. Figure 10.9 depicts this model. However, regarding rules, things are a little more complicated because they are encoded in a language that remains to be defined, for example from an extension of RuleML [BOL 10]. Indeed, for the design of the code of these rules, knowledge must be included from not only geographic objects and the relationships between them, but also ontologies (especially for the types of objects), the names of places and mathematical models as shown in Figure 10.10.

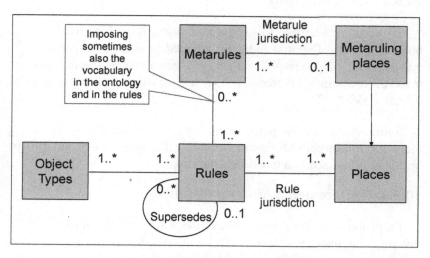

Figure 10.9. *Modeling geographic rules*

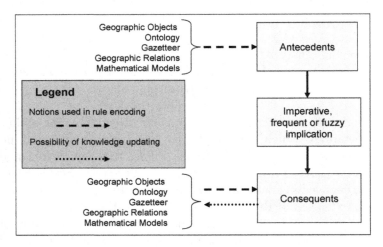

Figure 10.10. *Geographic knowledge involved in the geographic rules*

10.5.3. *Rules indexing*

To accelerate access to geographic rules, it will be necessary to build a system of spatial indexing adapted to this particular case. In the indexing of geographic objects, it seems that systems based on trees of rectangles (R-trees) [GUT 84] or their variants are most effective [VAN 02].

Similarly, to every place mentioned in each rule, a bounding rectangle minimum (Minimum Bounding Rectangle) will have to be built and organized in a tree-like structure.

10.5.4. *Requirements for a rule language*

From these observations, to describe a set of geographic rules, in this new language, two levels will be necessary, the rules themselves and their sets. As a first step to simplify the problem, it will not handle metarules and their consequences.

10.5.4.1. *Rule set level*

It will be important to include the elements common to all rules included in this set, that is the name of the set, language, jurisdiction, ontology and gazetteer. It could include references to other sets of rules provided that the language, ontology and gazetteer are compatible.

Then we will be given the rules themselves.

10.5.4.2. *Encoding rules*

As explained in Figure 10.10, three parts must be analyzed.

In the "Antecedent" part, it will be necessary to give the jurisdiction of the rule. This could be a place defined by a polygon with its coordinates, a toponym, or a Boolean combination of toponyms. This jurisdiction must be included within the jurisdiction of the rule set, for instance by a WHERE clause.

Then a list of relevant geographic objects and possibly Boolean conditions will follow.

The part "Implication" should indicate whether this rule is imperative, fuzzy, frequent or good practice.

As "Consequent", there may be the change of geographic objects, geometrically or semantically, and launching of actions.

In addition, for a rule, it would be interesting to add metadata and explanatory text. The role of metadata will primarily be to mention the origin of the rule, legal type, data mining, good practice, etc. as well as the date and the name of the editor. The explanatory text will be the text that will be printed when the user may wish to reconstruct the path of reasoning (traceability of the result).

10.6. Conclusion about applicative geographic rules

The purpose of this chapter was to illuminate the notion of geographic applicative rules by including examples. Unlike the rules of management in enterprises, we have tried to show the importance of space and the difficulties it could lead to. This importance is characterized by the necessity to consider IF-THEN-Zone rules, in addition to conventional IF-THEN-Fact and IF-THEN-Actions rules which can possibly also integrate geographic aspects.

Ultimately, the strong elements of the semantics of this type of rule were extracted, which allowed us to develop a first model. Now, we need to build an inference engine capable of integrating and reasoning with this type of semantics.

Therefore, it is possible to give a definition of a geographic rule that can be set as an imperative or modulated implication (frequent, desirable, etc.) involving either places or geographic objects, or both.

The other possibility will be to take account of temporal aspects to describe the rules of evolution of geographic objects such as shape changes (forests and deforestation, urban sprawl, dissemination, etc.). Also 3D should be included at term.

But first and foremost, it is necessary to continue this analysis in order to introduce other cases and thus enrich the semantics of the geographic rules. Then, it will be possible to propose a robust, consistent and effective formalism for representing the geographic rules and enabling them. We will also need to define the precise specifications of the actions to be undertaken for the treatment of the modulated implications.

11

Geovisualization and Chorems

The objective of this chapter will be to give some elements regarding the way to visualize geographic data and knowledge for decision-makers. Indeed, conventional cartography presents some limits and sometimes decision-makers require for novel types of visualization which can help them efficiently. Under the banner of visual analytics and geovisualization, new approaches are emerging, sometimes far from classical mapping.

What is a map? See Figure 11.1 for details. Several points of view can be defined. A map is a planar representation of the Earth with relative position of geographic features, sometimes accompanied with non-spatial attributes. This is a constructed model implying selection, down-scaling and generalization. It is also a graphic model usually using visual signs through a medium, at a given time and with specific goals. The previous considerations imply some choices made by the cartographer.

However, where is the problem? Why is conventional cartography criticized? What are the drawbacks? Among them, let us mention:

– since during centuries maps were not so easy to make, usually they have been illustrating too many details, maybe in a form of a set of maps such as atlases;

– among consequences, let us claim that they were hiding sometimes what is the more important for decision-makers;

– based on sensor networks, real time animation can be emphasized especially for dashboards;

– but the real objective, is it to describe reality or to emphasize the salient aspects?

In this chapter, after some remarks regarding graphics semiology, a few approaches in geovisualization will be presented together with chorems which are a way to schematize territories. To conclude this chapter, some aspects regarding geographic dashboards generated in real time will be given.

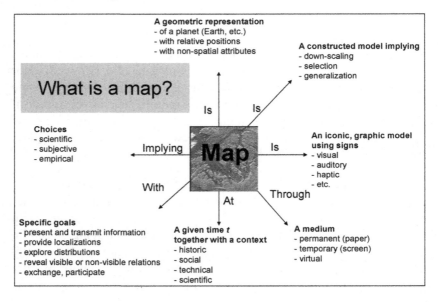

Figure 11.1. *What is a map?*

11.1. Graphics semiology

According to Bertin [BER 73], graphic semiology is "the set of rules of a graphic sign system for the transmission of information". Graphic semiology is so a system of signs that can be used to understand any graphics from the Highway Code to cartography; it is a discipline that concerns: the transcription of data into a sign in the graphics system, the processing of the data to show the information, and the design of images to provide this information.

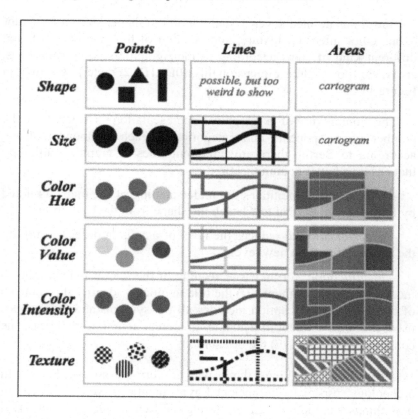

Figure 11.2. *Bertin's six fundamental visual variables, in relation with the type of graphic signs (or with the dimensionality of the graphic object) [BER 73]. For a color version of the figure, see www.iste.co.uk/laurini/geographic.zip*

·It uses the properties of the plan to show the relationships of similarity, order or proportionality between given datasets. In addition, a graphic system is composed by marks, also called "signs". Marks or signs are graphical objects that can be points, lines, areas, surfaces etc.; lines are visualized by signs of some thickness; areas have a length and width and are in a two-dimensional space; surfaces are areas in a three-dimensional space, but with no thickness, and volumes that have a length, a width and a depth. The described marks or signs are the basic units of graphics.

Bertin has defined seven visual variables (shape, size, color, color value, color intensity, texture, and position of the graphic sign), and different kind of signs according to their graphic sign (points, lines, areas, etc.), or to their dimensionality (0D, 1D, 2D, etc.) as shown in Figure 11.2.

The complexity of a graphical representation is related to the number of components and categories in each component. Then, according to Bertin, there are important rules in graphic semiology that ensure a better communication:

– *readability*: it facilitates the understanding of the map. As noted by Bertin it is important to "detach the shape from the background";

– *generalization*: it reduces the level of details in order to simplify the data to adjust to a new level;

– *identification*: for an easier reading of the map it is important to incorporate some elements: (*i*) the title to quickly identify the contents of the map; (*ii*) the caption includes all the symbols and color codes used; the scale concept of actual size; source; author: authenticate the contents of the map; (*iii*) the orientation.

Historically speaking, there was a very strong evolution. In archeology, especially in ancient Egypt, paintings were representing how things were known. Later, during the Renaissance, painters tried to show their reality as it was seen. Then the advent of photograph has completely reoriented painting. As a consequence, the key-idea of impressionism was to represent feelings. Concerning cartography, originally speaking, to show how to go from one place to another and to map what we know over the Earth. Now, the objective seems to show what could be important for decision-makers.

[BER 73] and overall [ITT 74] have constructed the basis for new representations. So now, cartography can be renovated.

11.2. From vision analytics to geovisualization

As cartography is to conventional way to visualize geographic information, now are emerging new techniques under the umbrella of visual analytics and geovisualization.

11.2.1. *Visual analytics – recommendations*

Geovisualization or "Geographic Visualization" concerns the visual representations of geospatial data and the use of cartographic techniques to support visual analytics. According to [THO 05], here are a few recommendations in research about visual analytics:

– build upon theoretical foundations of reasoning, sense-making, cognition, and perception to create visually enabled tools to support collaborative analytic reasoning about complex and dynamic problems;

– conduct research to address the challenges and seize the opportunities posed by the scale of the analytic problem. The issues of scale are manifested in many ways, including the complexity and urgency of the analytical task, the massive volume of diverse and dynamic data involved in the analysis, and challenges of collaborating among groups of people involved in analysis, prevention, and response efforts;

– create a science of visual representations based on cognitive and perceptual principles that can be deployed through engineered, reusable components. Visual representation principles must address all types of data, address scale and information complexity, enable knowledge discovery through information synthesis, and facilitate analytical reasoning;

– develop a new suite of visual paradigms that support the analytical reasoning process;

– develop a new science of interactions that supports the analytical reasoning process. This interaction science must provide a taxonomy of interaction techniques ranging from the low-level interactions to more complex interaction techniques and must address the challenge to scale across different types of display environments and tasks;

– develop both theory and practice for transforming data into new scalable representations that faithfully represent the content of the underlying data;

– create methods to synthesize information of different types and from different sources into a unified data representation so that analysts, first responders, and border personnel may focus on the meaning of the data;

– develop methods and principles for representing data quality, reliability, and certainty measures throughout the data transformation and analysis process;

– develop methodology and tools that enable the capture of the analytic assessment, decision recommendations, and first responder actions into information packages. These packages must be tailored for each intended receiver and situation and permit expansion to show supporting evidence as needed;

– develop technologies that enable analysts to communicate what they know through the use of appropriate visual metaphor and accepted principles of reasoning and graphic representation. Create techniques that enable effective use of limited, mobile forms of technologies to support situation assessment by first responders. Support the need for effective public alerts with the production of a basic handbook for common methods for communicating risks;

– create visual analytics data structures, intermediate representations, and outputs that support seamless integration of tools so that data requests and acquisition, visual analysis, note-taking, presentation composition, and dissemination all take place within a cohesive environment that supports around-the-clock operation and provides robust privacy and security control;

– develop an infrastructure to facilitate evaluation of new visual analytics technologies;

– create and use a common security and privacy infrastructure, with support for incorporating privacy-supporting technologies, such as data minimization and data anonymization;

– use a common component-based software development approach for visual analytics software to facilitate evaluation of research results in integrated prototypes and deployment of promising components in diverse operational environments;

– identify and publicize best practices for inserting visual analytics technologies into operational environments;

– form university-led centers of excellence as well as partnerships with government, industry, national laboratories, and selected international research entities to bring together the best talents to accomplish the visual analytics R&D agenda.

11.2.2. *Definition of geovisualization*

According to [MCE 97], Geovisualization, short term for "Geographic Visualization" can be defined as a set of tools and techniques to support geospatial data analysis through the use of interactive visualization. Like the related fields of scientific visualization and information visualization, geovisualization emphasizes information transmission. Geovisualization communicates geospatial information in ways that, combined with human understanding, allow data exploration and decision-making processes.

To summarize, geovisualization is an interesting and useful field of research for different reasons:

1) can reduce the time to search information, and support decision-making;

2) can enhance the recognition of patterns, relations, trends and critical points etc.;

3) can give a global vision of a situation, a phenomenon, etc.;

4) enables the use of human visual memory and the capability of perceptual processing of data;

5) permits a better interaction between user and the information system;

6) and can possibly lead to the discovery of new bunches of knowledge.

11.2.3. *Cartograms*

Cartograms are a first attempt to produce new maps in which territories are not represented in proportion of their areas but according to another variable. For example Figure 11.3 represents the New York City population. Another example is given in Figure 11.4 by depicting the results of the presidential elections in the US in 2004. In contrast with maps previously made, we claim that those cartograms better show the reality.

Figure 11.3. *Cartogram of New York City. Source: http://www. viewsoftheworld.net/?p=2071. For a color version of the figure, see www.iste.co.uk/laurini/geographic.zip*

Figure 11.4. *Results of the presidential elections in the US in 2004: a) Results by states; b) Circles proportional to population. Source http://www. goldensoftware.com/. Reproduced with permission. For a color version of the figure, see www.iste.co.uk/laurini/geographic.zip*

It is generally thought that cartograms distort reality. Indeed, by forcing some details, fraudulent maps can be created.

Figure 11.5. *DataAppeal Application showcasing datascape of CO_2 Levels, in Grenoble, France, rendered in light yellow spiky model. Reproduced with permission. For a color version of the figure, see www.iste.co.uk/laurini/ geographic.zip*

11.2.4. Examples in geovisualization

Abundant research have been carried out in geovisualization. Among them, Nadia Amoroso's works are of interest. For instance Figure 11.5 shows a new way to represent air pollution in the city of Grenoble, France, whereas Figure 11.6 represents population. These examples[1] illustrate news directions which can be useful for local decision-makers.

---

1 Both Figures 11.5 and 11.6 come from the following web site http://archinect.com/ features/article/71075299/working-out-of-the-box-nadia-amoroso.

Figure 11.6. *Datascape of NYC's population, rendered in a light blue bubble model representation. Over 20K data-points rendered instantly. Reproduced with permission. For a color version of the figure, see www.iste.co.uk/laurini/ geographic.zip*

11.2.5. *How to lie with maps?*

Now that it is clear that some deformations should be made to enlighten local authority decision-makers, it is nevertheless important not to produce fraudulent maps. In 1991, Monmonnier [MON 91] produced a very exciting book entitled *How to lie with map?* In this book, he showed many examples of how reality can be distorted or disguised. Figure 11.7 give an example of a funny busline. In Figure 11.7(a), one can see the "real" map whereas in Figure 11.7(b) lays a commercial announcement. In this announcement, it is easy to see that the busline look quicker: it is a fraudulent map. However in Figure 11.8(b), the caricature of a road network (Figure 11.8(a)) is more easily understandable than a map nearer to the reality showing too many details hiding the more important issues.

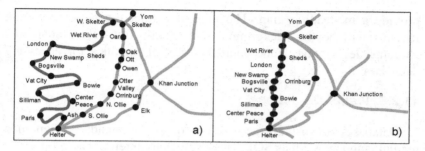

Figure 11.7. *A funny busline from Monmonier: a) a "real" map;*
b) a fraudulent map. Redesigned from [MON 91]

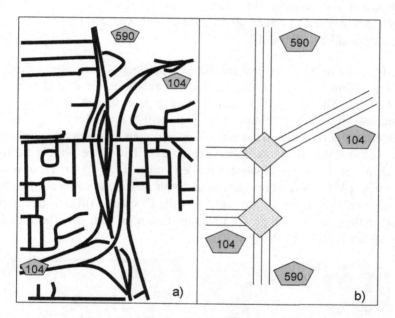

Figure 11.8. *A road network and its caricature: a) a "real" map,*
not so easy to understand; b) a caricature more easily
understandable. Redesigned from [MON 91]

From those previous examples, we can learn that the borderline
between a fraudulent map and a caricature showing the more
important is very tiny. As a concluding remark, let us say that
sometimes it is not interesting to show all details, but one does not

provide a misleading map. The idea to extract the more important features will be the background of chorems. But before analyzing chorems, let us give some characteristics of graphics for accessing databases.

11.2.6. *Visual DB access*

Databases are usually accessed through commands written in a textual language such as SQL. However, Ben Shneiderman proposed a mantra for designing human interfaces: "Overview, zoom and filter, details on demand" [SHN 97a, SHN 07b], that is macroscopic versus microscopic approach. This mantra was recently refined in "Analyse First, Show the Important, Zoom, Filter and Analyse Further, Details on Demand" by [KEI 06b].

Indeed, for conventional databases, approaches such as starfield or space filling treemaps were created for relational or object-oriented databases. The starfield system is targeted to layout instances of a database object or a relation into a screen: a procedure is given for selecting the two axes from attributes, and then a third axis is selected for colours; the result is called a starfield. The more widely known example is the starfield system made (Figure 11.9(a)) for Hollywood movies [AHL 94]. For databases with different objects, another metaphor is used based on so-called space filling treemaps; personally, we would prefer to name this approach the "bookshelf" metaphor (Figure 11.9(b)).

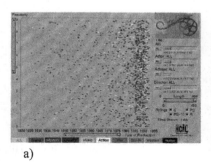 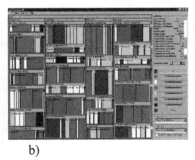

a)   b)

Figure 11.9. *Ben Shneiderman system for DB accesses: a) Starfield; b) Space-filling Treemaps. Reproduced with permission. For a color version of the figure, see www.iste.co.uk/laurini/geographic.zip*

11.3. Chorems

Chorems were created in 1980 by Pr. Roger Brunet, a French geographer [BRU 80] as a schematic representation of a territory. This word comes from the Greek χώρα which means space or territory. It is not a raw simplification of the reality, but rather aims to represent the whole complexity with simple geometric shapes. Even if it looks a simplification, the chorem tries to represent the structure and the evolution of a territory with a rigorous manner.

The basis of a chorem is in general a geometric shape in which some other shapes symbolize the past and current mechanisms. Brunet has proposed a table of 28 elementary chorems, each of them representing an elementary spatial configuration, and so allowing them to represent various spatial phenomena at different scales. According to Brunet, chorems are a tool among others to model the reality, but it is a very precious tool not only as a visual system, but also as a spatial analysis tool.

As an introductive example, let us mention the water problem in Brazil as depicted in Figure 11.10. In Figure 11.10(a), there is a conventional map of rivers. In Brazil, as usual, at this scale, only the main rivers are mapped, therefore this map will not be very useful for decision making. However, nothing indicates what and where the problems are. However, Figure 11.10(b) gives a chorem map of the situation according to the caption given Figure 11.10(c). We can see more accurately where the humid and dry zones are, where water is missing, where dykes are located, where water is under demand and so on. As a conclusion, this kind of drawing is much more informational to any decision-maker than the conventional hydrographic map.

After a more accurate definition of chorems, we will describe how chorem map can assist decision-makers.

Then, we will present how chorem map can help create a novel progressive entry system for geographic databases following Ben Shneiderman's mantra. The ultimate goal will be to create the main concepts in order to implement a decision support system which can be schematized as follows (Figure 11.11): starting from the real world, a geographic database or datawarehouse will be constructed and populated. Then by applying appropriate spatial data mining and filtering techniques, geographic knowledge will be discovered and then visualized as chorems [LOP 06].

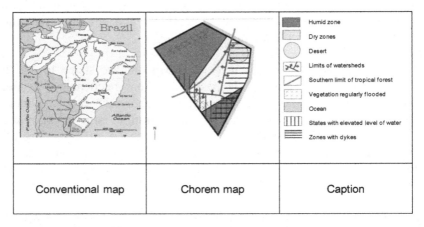

Figure 11.10. *Comparing a conventional map of rivers in Brazil and a chorem map emphasizing the water problem in this country². For a color version of the figure, see www.iste.co.uk/laurini/geographic.zip*

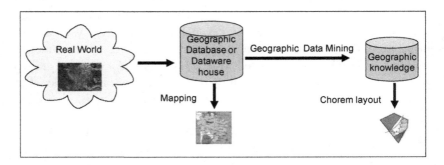

Figure 11.11. *Main steps of the proposed methodology*

---

2 This example is drawn from the web site http://histoire-geographie.ac-bordeaux.fr/ espace eleve/bresil/eau/eau.htm [LAF 05].

What are chorems? How can they help decision-making? Those are important questions that we will try to answer in the subsequent sections. In our case, we will give the following definition: "chorems are a type of map generated from both geometric and semantic points of views".

11.3.1. *Elementary chorems*

As initially stated, a table of 28 chorems was created by Brunet; Table 11.1 gives the Brunet's basic chorem vocabulary.

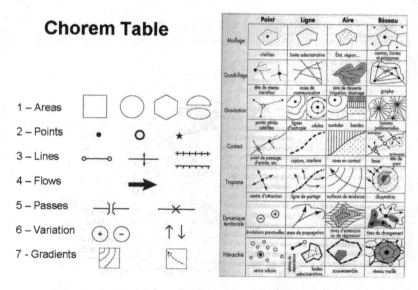

Table 11.1. *Table of Brunet's chorems [BRU 86]*

Finally chorems can be defined as a visual vocabulary allowing the description of salient characteristics and problems of a territory. From our point of view, chorems are a good basis for decision making, because they emphasize the more significant aspects, leaving out secondary problems. Let us say that when it is necessary to understand the structure of a territory, a one-to-one map is not useful, whereas a small schema can be a more useful scale. So chorem maps are a key-tool to schematize a territory, which allows decision-makers or politicians to get a clearer view of the situation. Amongst other applications, let us mention the schematic visualization for:

– salient political, economic and demographic problems;

– salient features in environment and climatology;

– interesting places in geomarketing;

– main evolution in epidemiology;

– natural and technological risks or disasters;

– etc.

Regarding other applications, let us mention a very interesting approach to use chorems (called here choremes) for way finding by [KLI 05] and [REI 09].

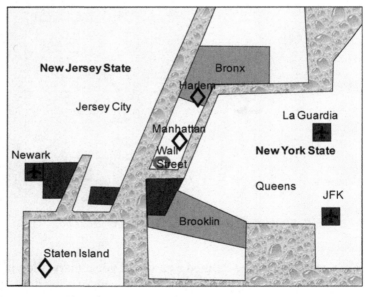

Figure 11.12. *Example of chorematic representation of New-York City*[3]

---

3 Translated from https://cahiersdhistoire.net/varia/archives-cartographiques/schema-de-la-ville-de-new-york/.

An example is given Figure 11.12 for the city of New York. Among the salient aspects, let us mention: the importance of Wall Street as the world financial center, the "barrier" between Manhattan and Harlem, the role of technological poles, the airports, etc.

Another example is extracted from [CHE 15a] showing the internal migrations in Tunisia emphasizing the attraction of the capital city of Tunis (see Figure 11.13).

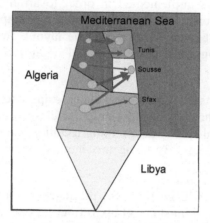

Figure 11.13. *Chorem showing internal migrations in Tunisa. Adapted from [CHE 15a]*

11.3.2. *Approaches, manual versus automatic?*

Chorems can be used for modeling several territories, for instance to extract salient features in:

– relief and climatology;

– ecosystems, environment;

– history, population and demography;

– rural and urban dynamics;

– communication networks;

– economy and international relationships.

As far as we know, no automatic approaches seem to exist. As a consequence, chorems appear as geographic knowledge which must be visualized. In other words, we can define now chorems as a kind of visual geographic knowledge.

To be more precise, at the discovery level of chorems, geographic knowledge is chased. Once it is discovered, it must be not exactly mapped, but rather visualized. That is to say that some layout procedure must be implemented in order to arrange the appropriate elementary chorems. During this step, perhaps some spatial knowledge must be used to perform the layout, for instance by not overlapping cartographic objects, spatial organization of elementary diagrams, etc.

The big advantage of the Pech-Palacio's approach is to be a global approach, that is he considers not only one class of geographic objects as done by several spatial data mining techniques, but rather all kinds of geographic objects. By using this method, geographic knowledge will be discovered.

The next step will be to select the more salient knowledge. For that, a list of criteria must be set up to filter all that knowledge chunks which can be represented as chorems. We will use two criteria, first on importance values and cause–effect relationships.

11.3.2.1. Importance based on values

The first criterion will deal will the distribution of geographic objects or characteristics [BOU 17]. A solution is to define importance from a Gaussian distribution. Indeed:

– a value with a deviation from the average lower than the standard deviation is referred as trivial;

– a value with a deviation between once and twice standard deviations is considered moderately important;

– a value with a deviation between two and three times standard deviations begins to be remarkable;

– a value with a deviation between three and four times standard deviation is considered to be exceptional;

– a value with a difference of more than four times the standard deviations is described as historic and very rare.

Based on this consideration, we can define an importance function β as the absolute value of the ratio between the average and the standard deviation. According to β we can conclude (Figure 11.14):

1) if $0 \leq \beta \leq 1$ Then the value is of a trivial importance;

2) if $1 \leq \beta \leq 2$ Then the value is of moderately importance;

3) if $2 \leq \beta \leq 3$ Then the value is of remarkable importance;

4) if $3 \leq \beta \leq 4$ Then the value is of exceptional importance;

5) if $\beta \leq 5$ Then the value is of historical importance and very rare.

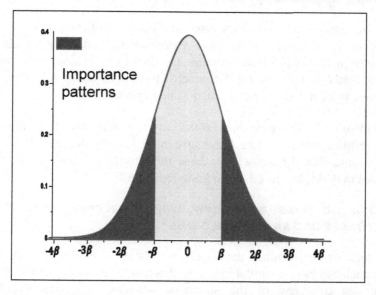

Figure 11.14. *Importance based on Gaussian standard deviation*

This method can be used within the territory under consideration, but also with a larger scope. If we refer back to the New York City example, the importance of the financial district cannot be detected within the city (because there is only one financial district) but rather by study all financial centers all over the world.

As this method is interesting to identify, salient values of geographic objects with the same ontological type, it does indicate how to rank salient values between objects of different types. However, nothing prevents to mix different β's coming from different sources, to rank them and consider their greatest values.

11.3.2.2. Importance based on causality

For instance in Latin America, consider the border between Argentina and Chile. The loose ribbon shape of Chile is explained by the presence of Andes or Andean Mountains (Spanish: Cordillera de los Andes) and the position of the routes going from one country to the other are governed by passes.

So, suppose we have discovered an important phenomenon, say A. Suppose in addition that we have a rule stating that $B \Rightarrow A$, and we know that B exists in our system. By backward chaining, we must claim that B is important. Of course, if the causal chain is longer, the origin can be detected and claimed to be important.

However, suppose we have $B \vee C \Rightarrow A$. In this case, there are two possibilities, either to state B as cause or C. The problem is that we are never sure that the knowledge base is complete so that some other causes may not be stored in the knowledge base.

11.3.3. Chorems as a new way to access geographic databases and knowledge bases

Back on geographic databases and datawarehouses, the chorem approach can have a similar target. In this case, the chorem of a territory gives an overview of the situation, whereas, following the Ben Shneiderman's mantra, the "details-on-demand" step can be represented by a detailed mapping. By "zooming and filtering", we can gracefully and gradually reduce the search space. Here zooming will mean using different geographic scales or thematic disaggregation, whereas filtering reflects conditions and criteria (geographic and semantic zooming). By zooming and filtering, a sort of sub-chorem can be defined. By sub-chorem, we mean a chorem made for a smaller territory. For instance, the first step can be a chorem for a whole country, then chorems for regions and so on.

In other words, chorems can be seen as a new way to enter geographic databases. Table 11.2 gives a comparison between the use for conventional databases, and the approach to geographic databases and datawarehouses. And Figure 11.15 schematized the comparison of various styles of database entry systems.

So, a new way (Figure 11.16) of entering a geographic database can be sketched:

– at the opening, a global chorem map can be displayed;

– then by semantic and geographic filtering some sub-chorem maps can be visualized;

– finally, the final query answer (map or table) can be displayed.

Ben Shneiderman's mantra	Conventional databases	Chorem-based approach
Starting point	Relational or object-oriented database of an organization	Any kind of data which can be useful
1 – Overview	Generally the "overview" is visually presented by means of starfield or space filling treemaps; they are both structure- and content-oriented.	The territory-level chorem can give an overview, perhaps more linked to problems than to data contents.
2 – Zoom and filter	Criteria can be used to reduce the search space.	The territory can perhaps be split into different zones, each of them with a sub-chorem (geographic zoom). A second way can be to reduce the number of topics (semantic zoom)
3 – Details on demand	The final step delivers what could be necessary for the user, usually as a table.	Here both tables and maps can be the final steps, depending on the user's needs.

Table 11.2. *Comparing accesses to conventional and geographic databases*

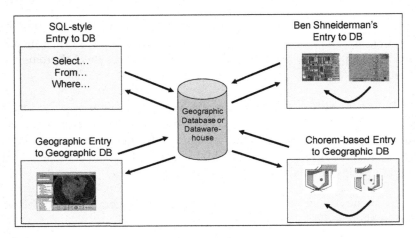

Figure 11.15. *Comparing various styles of database entry systems*

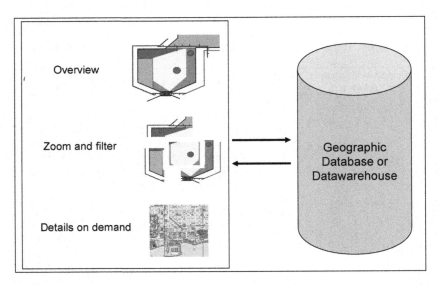

Figure 11.16. *Chorem-style entry to geographic databases*

Another potential use of chorems issued from data mining [DEL 07] are to generate visual summaries of geographic databases. Hence, by analyzing them, some salient issues can be discovered and laid out so to represent "what is important in a geographic database".

In other words, geographic knowledge can be represented visually and chorems could be an elegant way not only to access geographic knowledge bases but also as output.

11.3.4. *Final remarks regarding chorems*

The goal of this section was not to present any result, but rather to present some landmarks for a research plan in geographic decision-making based on chorems. Three main aspects were emphasized:

– chorems are a map generalization both geometrically and semantically;

– chorems can be a way to represent geographic knowledge, as visual summaries coming from spatial data mining;

– chorems can be a new way to enter geographic databases as a global vision.

In any case, this research plan must be detailed, concepts must be clarified, and experimentations must be made through several prototypes.

11.4. Dashboards for smart cities

Visualization, and in particular impromptu and real time geovisualization, is very important for the governance of smart cities and smart territories based on data coming from sensor networks. Applications can be found for traffic management, pollution control, meteorology, disaster management, etc. Two types of dashboards can be considered:

– multi-thematic dashboards, that is in which several indicators are shown simultaneously, maybe thousands on maps;

– chorem-based dashboards, that is in which only the salient aspects, usually named hotspots are recognized and laid-out.

Only the second will be rapidly described based on the Ben Shneiderman's mantra. In the overview, all salient aspects discovered in real time are shown. Then the step zooming-and-filter will gradually lead the user to the details relative to the concerned hotspots.

Some prototypic examples are given in the literature[4].

11.5. Conclusions

The goal of this chapter was to give new tracks not exactly for renovating cartography, but overall to present new solutions in geovisualization in order to help decision-makers in charge of the planning and management of smart cities and smart territories.

In our view, as explained in the previous sections, chorems seem to be a very promising way to discover and emphasize salient issues hidden in a territory as they can be also used as a visual access technique and as an output from any geographic inference engine.

---

4 See http://urbact.eu/steering-real-time-city-through-urban-big-data-and-city-dashboards-0 and http://cityofthefuture-upm.com/first-data-through-the-upms-smart-city-dashboard/.

GKS: Querying and Interoperability

Now that the majority of aspects of modeling geographic knowledge systems has been outlined, it could be important to use them. Since geographic inference engines do not yet exist, let us examine two facets, how could one query them and organize some interoperability?

12.1. Geographic queries

Beyond geographic databases in which extensions of SQL were created to deal with geographic aspects, it could be of interest to study novel querying systems for instance as extensions of geovisualization as explained in the previous chapter.

12.1.1. *Textual queries*

If you consider the various components of a geographic knowledge system, it is obvious that a querying system should be able to help writing easily geographic queries dealing sometimes with geographic relations, structures, gazetteer and ontologies. However, such a task could be the basis of a new research initiative.

In contrast, this book will not pay more intention to this issue, but only visual querying will be developed in the next section.

12.1.2. *Visual languages and queries*

To design a visual language, one has to define the vocabulary and the grammar of said language in order to define statements (= knowledge) and define interrogations (queries), and the context of interpretation. Among the generic characteristics, a visual language must be universal, that is, everybody must understand it; in other words, icons must come from a fully-agreed visual ontology, and object icons must be known by anyone. In addition its expressive power must be as large as possible:

a) *Vocabulary*

Not only object types must be represented by an icon but also object classes themselves. As type icons [BON 99] (Figure 12.1) must be defined in the ontology, object icons (e.g. a city, a river) must be defined in a visual gazetteer (Figure 12.2). Some cities have emblems or seals which can be used as countries have flags. As the State of Mississippi has a flag and a seal, the Mississippi river has no official symbol (as far as we know). The consequence is that if somebody creates an icon for the Mississippi river, a lot of people will have difficulties understanding either the statement or question its meaning.

The main difficulty in defining geographic object icons is that in cartographic legends, they have various representations. For instance, restaurants and museums have dozens of visual representations.

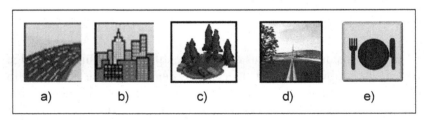

Figure 12.1. *Examples of geographic object type icons: a) river; b) city; c) lake; d) route; e) restaurant*

Concerning mathematical vocabulary, let us propose to continue using the usual symbols.

Figure 12.2. *Examples of geographic object icons issued from a visual gazetteer, seals of cities of Boston, Baltimore and Philadelphia, bottom of Geneva and Marseilles, then at right flags for the USA, Maryland State and France*

b) *Sentences*

Three types of sentences must be defined, statements, constraints, rules and queries. Statements will correspond to geographic facts, clusters and flows. Rules must be defined by using the => or ⇒ symbol. And queries by using Spanish interrogation marks (¿ ?) which can be used as parentheses.

c) *Contexts of interpretation*

In fact, four types of interpretation spaces are possible, identified by interpretation icons (Figure 12.3):

– *Cartographic space* which corresponds to conventional cartography with an arrow to North and a scale; the horizontal axis represents eastings; this context is identified by the North arrow icon; according to scale, it can be based on projections or being spherical; It has two alternatives, the planar one (iconized by a square) and the global one (circle);

– *Topological space* in which only cardinal directions have no importance, but the importance is given to the respective positioning of geographic objects; the horizontal and vertical axes have no meaning; this context is identified by an "overlap relation";

– *Time line* in which the horizontal axis represents time; this context is identified by a clock icon; remind that this interpretation context is outside the scope of this paper;

– *Chorematic space* showing a geometric and semantic generalization of a territory; this case is represented by a hexagon because sometimes France is caricaturized as a hexagon.

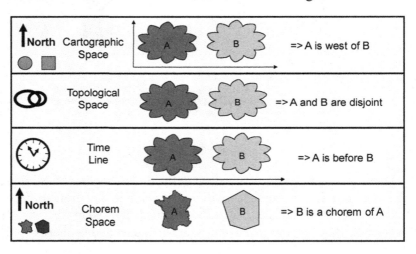

Figure 12.3. *The contexts of interpretation, cartographic space, topological space, time line and chorematic space*

d) *Examples*

Figure 12.4 is taken in the planar cartographic context. Figure 12.4(a) illustrates a fact that the city of Baltimore is located south of Boston, whereas Figure 12.4(b) represents a query in order to know whether Baltimore is located south of Boston.

Figure 12.5 illustrates a more complex topological query in order to get rivers crossing the State of Maryland. Since the scale is mentioned, it means that one is only interested by rivers wider than

100 meters. Figure 12.6 represents a topological constraint concerning the cities of Marseilles (*Covers*) and Geneva (*Touches*) with France.

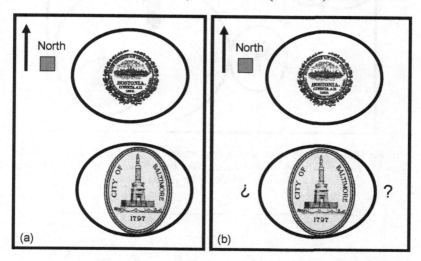

Figure 12.4. *A fact and a query in a geometric context: a) representing the fact or statement that Baltimore is south of Boston; b) representing a query in order to know whether Baltimore is located south of Boston*

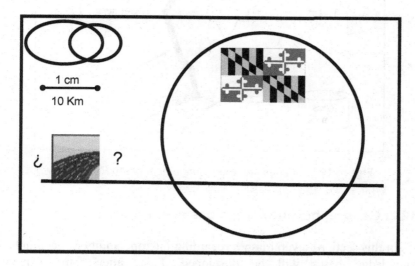

Figure 12.5. *A topological query to get the list of rivers crossing the State of Maryland*

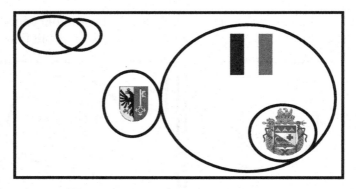

Figure 12.6. *Example of visual representation of constraint stating that the city of Marseilles must be always inside (*Touches *relation) the French territory, and Geneva outside (*Covers*)*

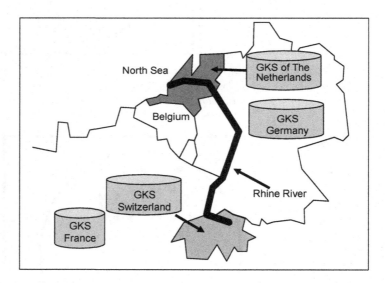

Figure 12.7. *Example for the management of the Rhine River*

12.2. Geographic knowledge bases interoperability

In this section, several issues regarding interoperability of geographic knowledge bases will be developed. First, after some general considerations regarding interoperability, interoperability based on ontologies and views will be explained. Then the problem of federating several geographic knowledge bases will be rapidly developed.

Indeed, some important problems in smart governance cover several jurisdictions. In other words, it is sometimes necessary to use several knowledge bases, each belonging to a different jurisdiction. In Figure 12.7, an example of the management of the Rhine River implies to use several knowledge bases covering Switzerland, France, Germany, The Netherlands and perhaps Belgium.

12.2.1. *Generalities about interoperability*

Interoperability is a very important concept allowing several databases to cooperate. It can be defined as the ability of a computer system to run application programs from different vendors, and to interact with other computers across local or wide-area networks regardless of their physical architecture and operating systems.

For territorial intelligence, two main cases must be distinguished:

– when several geographic knowledge bases are covering the same territory (Figure 12.8(a)), the main problem is that same objects can have different geometric representations and semantic attributes; this problem is usually solved by means of ontologies (see Chapter 6);

– and when two or more data knowledge bases are covering different territories (Figure 12.8(b)) but having some borders in common. This phenomenon is called cross-border interoperability.

In this book, which is dedicated to geographic knowledge infrastructure, the main problem concerns the construction of a new unique infrastructure based on several existing GKS. By comparing this problem to databases, it can be called GKS federation. So now, let us examine various facets to federate two different GKS. According to Date [DAT 87], a well done federation must follow 12 rules, namely.

1) *Local autonomy*. Local data are owned and managed locally, with local accountability and security. No site depends on another for successful functioning.

2) *No reliance on a central site.* All sites are equal, and none relies on a master site for processing or communications.

3) *Continuous operation.* Installations at one site do not affect operations at another. There should never be a need for a planned shutdown. Adding or deleting installations should not affect existing programs or activities. Likewise, portions of database should be able to be created and destroyed without stopping any component.

4) *Location independence (transparency).* Users do not have to know where data are physically stored. They act as if all data are stored locally.

5) *Fragmentation independence (transparency).* Relations between data elements can be fragmented for physical storage, but users are able to act as if data was not fragmented.

6) *Replication independence.* Relations and fragments can be represented at the physical level by multiple, distinct, stored copies or replicas at distinct sites, transparent to the user.

7) *Distributed query processing.* Local computer and input/output activity occurs at multiple sites, with data communications between the sites. Both local and global optimization of query processing are supported. That is, the system finds the cheapest way to answer a query that involves accessing several databases.

8) *Distributed transaction management.* Single transactions are able to execute code at multiple sites, causing updates at multiple sites.

9) *Hardware independence.* Distributed database systems are able to run on different kinds of hardware with all machines participating as equal partners where appropriate.

10) *Operating system independence.* Distributed database systems are able to run under different operating systems.

11) *Network independence.* Distributed database systems are able to work with different communications networks.

12) *Database independence.* Distributed database systems are able to be built with different kinds of databases, provided they have the same interfaces.

By looking at those rules, we can see that they can be also used for knowledge bases and typically geographic knowledge bases. Amongst them, the more important are the three first rules, namely autonomous of each GKS, no central site and continuity of operations.

For conventional databases, two types of fragmentations of tables were usually distinguished, that is vertical and horizontal. However, for geographic databases, this distinction is not well adapted. The following is more adequate, layer fragmentation (Figure 12.8(a)) and zonal fragmentation (Figure 12.8(b)).

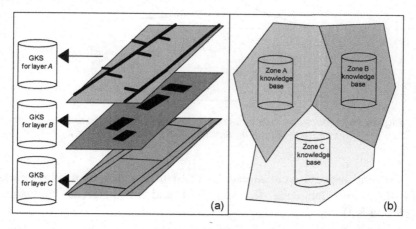

Figure 12.8. *Different types of geographic interoperability: a) various GKS covering the same territory; b) cross-border interoperability*

Both layer and zonal fragmentations can be solved by using ontologies.

12.2.2. *Interoperability of databases based on ontologies*

In the database community, ontologies were created to solve interoperability problems between different databases; in this context, each database (*A* and *B*) must have its own ontology (usually a sort of re-writing its conceptual model) called local ontology; and to communicate, a domain ontology (Figure 12.9) must be used to link concepts. For each concept of the local ontology, one or more concepts of the domain ontology must match. A special program called a mediator is in charge of this translation; in a sense (A➔B) it translates the query A into the B language using the domain ontology; then the query is made against B which gives a result. This result in the B language is again transformed by the mediator to deliver the answer in the A language.

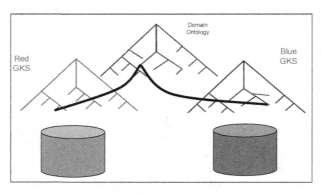

Figure 12.9. *Mechanism of local ontologies using a domain ontology with some equivalence relations between local ontologies*

Several aspects were already studied in the fusion of ontologies, in the same language (section 6.6) and in different languages (section 8.4).

However, federating geographic knowledge bases present particular characteristics once solved the problem of coordinates, the problems linked to tessellations, networks, in order to offer decision-makers a common vision in the spirit of Prolegomenon 12. The sub-problems regarding the federation of knowledge components are as follows (among others):

– federating the geographic objects imply to confer them global identifiers, consistent types and similar geometric representations; in addition, some objects (such as roads or rivers) artificially cut by jurisdiction borders must be reconstituted;

– federating relations, ontologies, gazetteers imply the use of a common human language to describe object characteristics and placenames;

– federating external knowledge is not so easy because a subset of the outside knowledge can also be present in the neighboring knowledge, maybe with different representation; in addition a global crown knowledge must be defined and integrated.

12.2.3. Federating tessellations

Practically, any knowledge base covers a jurisdiction, perhaps with external knowledge. In other words, it means that the concerned territory can be a piece of a higher tessellation and "brother" of neighborhood territories. From a practical point of view, since there are discrepancies at borders, the problem is similar to the matching of loose tessellations as explained in section 7.3.

12.2.4. Federating networks

The problem of federating networks is more complex. Indeed, sometimes a river or a road can delimitate the border or they can cross the border. Following the graph terminology, when they cross the border, one has a node, and when they are along a border, the border is an edge. But due to measuring errors, this is a little more complex. Let us study both of them. Do not forget that we must consider ribbon networks not only for road and rivers, but also for railways, power or telecommunications cables, perhaps for sewerages, etc.

a) *Federating nodes*

Figure 12.10 illustrates a network example in which two neighboring geographic repositories are present, obviously with geometric discrepancies [LAU 98]. Figure 12.10(a) shows two geographic repositories before integration; Figure 12.10(b) shows the results of cartographic integration (maps look good); a successive step is not mentioned in this figure is object integration in which two objects (for instance, a road, a river) which were artificially cut into two pieces, fusion, that is same identifier.

Then Figure 12.10(c) shows the last step, graph integration. Before integration road graphs are not connected, but in order to allow graph reasoning, for instance minimum path algorithm across several repositories, graphs must be connected; in this case a node must be created in which a first edge belongs to the first repository, and the second edge to the second one.

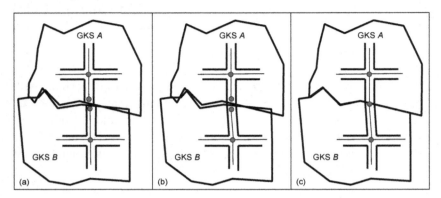

Figure 12.10. *Consequences of cross-boundary interoperability: a) Before integration; b) Cartographic integration; c) Graph-reasoning integration*

In other words, before integration there was a set of two non-connected ribbons and, at the end, the concerned ribbons are reduced to a unique ribbon, where a unique graph is constructed. As a consequence, three levels of interoperability can be distinguished:

1) *Geometric continuity*: by only force-fitting bordering points, the resulting map looks good; this is only a very low level of interoperability only for cartographic purposes;

2) *Semantic continuity*: reconstruction of geographic objects partly cut by knowledge bases (for instance roads, rivers) in order to provide them with the same toponym; this is really the preliminary level for seamless querying;

3) *Topological continuity* for ribbon networks: indeed, before integration, one has two disconnected graphs, and after the two graphs are connected to form a single network against which algorithms such as minimum path can be launched; this is the complete level of interoperability since all applications can be run.

Therefore, it is compulsory to provide necessary tools for both creating cross-boundary edges, and launching graph algorithms without blockage, not only for roads, but for any kind of networks as previously mentioned.

b) *Federating edges*

Often rivers constitute a natural borders between two countries, states, etc., a bank owned by one country and the other by the other country. So we deal with two GKS, namely A and B having such context, and several situations are possible in order to reconstitute the river ribbon. Let us examine all of them depending how both river banks are described: (*i*) the corresponding bank is not stored, (*ii*) a base stores its own corresponding bank, (*iii*) a base stored both banks according to Principle #11 as external knowledge. By combining those situations, several cases can hold (Table 12.1).

Cases	GKS *A*	GKS *B*	Actions to do
a		No	Keep *A* bank and reconstitute *B* bank
b			Keep both *A* and *B* bank, but check whether river width is plausible
c			Keep *B* borders and check whether *A* border is homologous of the *B* westernmost border
d		No	Adopt *A* borders
e			Check whether both borders are homologous. Take the more recent.

Table 12.1. *Showing the different cases to reconstitute river ribbon*

From a practical point of view, two possibilities can be envisaged. The first one is to try to automatically reconstitute the river banks, perhaps by designing rules based on Table 12.1; for instance by creating a decision table. For the second, since some difficult choices must be carried out taking quality into account, a sort of border editor must be designed and the task is assigned to some experts. In any case, the results must supersede existing corresponding knowledge chunks. Once this is done, we have to check whether tributaries are all connected.

Once this step is fulfilled, one needs to deal with networks. Before the federation, there were two networks of rivers, one for each GKS. Now the problem is to construct a unique network starting from the previous, together with the newly-reconstituted river ribbon.

12.2.5. *Federating GKS*

However, the previously described operations, namely federating geographic objects and structures, are not the only ones we have to carry out to federate two or several GKS. Indeed, the following additional tasks must be done:

– integrating relations; an example was given in section 8.5.3;

– integrating ontologies; examples were provided in the same language (section 6.6) and in different languages (section 8.4);

– integrating gazetteers and rules; here more research must be done to solve those problems;

– integrating physico-mathematical models; in this case, since they are encapsulated, federation looks easy;

– integration of external knowledge present two facets; indeed some chunks covered by other GKS were already integrated, but additional outside knowledge must be provided.

12.3. Conclusion

In this chapter only two facets in running GKS were dealt with, namely querying and interoperability. Many others should be worked on, for instance for geographic knowledge indexing.

Regarding querying, beyond textual querying, visual querying can be a good candidate for presenting queries, especially in connection with geovisualization in which some modes of interaction can be created.

The problem of interoperability is a very complex problem. In a previous talk, I have declared that "Interoperability is a dream for users, but a nightmare for systems developers". Here only a few directions are provided.

Now, let us examine how geographic bases can be used for governance both for smart cities and for territorial intelligence.

13

Conclusion: Knowledge as Infrastructure for Smart Governance

Now that structuring elements are defined and given, the problem is examining how geographic knowledge engineering can be integrated into territorial intelligence and smart cities, that is into the smart governance of cities, regions, etc. As previously mentioned, for years, companies have been using the so-called Business Intelligence software into practice with its extensions such as geospatial business intelligence. But in reality, there is an important gap between geospatial intelligence and territorial intelligence.

The aim of this chapter will be to explain the differences and show how territorial intelligence is more targeted at the government of local authorities. After the presentation of the main concepts in business intelligence, the key-elements of geospatial intelligence will be examined. Then, by sketching smart governance based on Knowledge Infrastructure, I will try to show how those concepts can renovate urban and environmental planning.

13.1. Business Intelligence

Various definitions of Business Intelligence (BI) have been given. For instance, according to Vangie Beal[1], "Business intelligence (BI)

1 http://www.webopedia.com/TERM/B/Business_Intelligence.html.

represents the tools and systems that play a key role in the strategic planning process within a corporation. These BI systems allow a company to gather, store, access and analyze corporate data to aid decision-making. Generally these systems will illustrate business intelligence in the areas of customer profiling, customer support, market research, market segmentation, product profitability, statistical analysis, inventory and distribution analysis to name a few". In other words, Business Intelligence is essentially a set of computer tools to help company executives make better decisions.

According to [BAD 09], business applications rely on a complex architecture of software that is usually composed of (see Figure 13.1):

– an extract/transform/load (ETL) tool to extract data from different heterogeneous sources, provide integration and data cleansing according to a target schema or data structure, and load the data in a data warehouse;

– a data warehouse which stores the organization's historical data for analysis purposes;

– an online analytical processing (OLAP) server which enables the rapid and flexible exploration and analysis of the large amount of data stored in the data warehouse;

– on the client side, some reporting tools, dashboards and/or different OLAP clients to display information in a graphical and summarized form to decision makers and managers. These tools are capable of exploring data interactively and supporting the analysis process;

– optionally, some data mining tools to automatically retrieve trends, patterns and phenomena in the data.

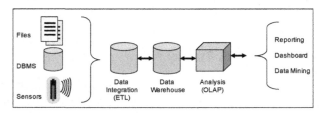

Figure 13.1. *Typical infrastructure on which BI applications rely, according to [BAD 09]*

13.2. GeoSpatial Business Intelligence or geo-intelligence

According to [BAD 09], "classical BI tools are often unable to handle the spatial dimension of data or only provide a very basic support. Some phenomena can only be adequately observed and interpreted by representing them on a map. This is especially true when you want to observe the spatial distribution of a phenomenon or its spatiotemporal evolution" (see Figure 13.2).

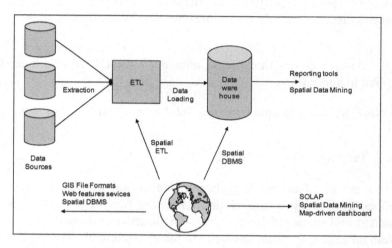

Figure 13.2. *Integrating the spatial component into a classical Business Intelligence infrastructure [BAD 09]*

For this, the existing tools are GeoKettle[2] and GeoMondrian[3]. GeoKettle is a powerful, metadata-driven Spatial ETL tool dedicated to the integration of different spatial data sources for building and updating geospatial data warehouses. GeoKettle enables the extraction of data from data sources, the transformation of data in order to correct errors, do some data cleansing, change the data structure and

---

2 http://www.spatialytics.org/projects/geokettle/.
3 http://www.spatialytics.org/projects/geomondrian/.

make them compliant to defined standards. In addition, it allows the loading of transformed data into a target DataBase Management System (DBMS) in OLTP or OLAP/SOLAP mode, GIS file or Geospatial Web Service. GeoKettle is a spatially-enabled version of the generic ETL tool Kettle (Pentaho[4] Data Integration).

The main criticism is that the expression "Business Intelligence" is not adequately applied to those tools. In our understanding, tools based on applied artificial intelligence should be designed and implemented not only for companies but also by local authorities.

It is of interest to mention [JOH 15] in which the authors propose a SOLAP tool linked with chorems, named ChoremOLAP which can be applied to agriculture.

Now, let us reexamine territorial intelligence and its aftermath.

13.3. Territorial intelligence

As explained in the first chapter, "territorial intelligence can be defined as an approach regulating a territory (maybe a city) which is planned and managed by the cross-fertilization of human collective intelligence and artificial intelligence for its sustainable development". The consequence of this definition is that we have to study several issues among others, such as the way to involve people and the way to involve knowledge technologies. Let us begin with people and public participation.

13.3.1. *Public participation*

The problem of the various degrees of which to involve people in urban planning is very old [LAU 01]. In 1969, Arnstein [ARN 69] proposed the first ladder for public participation with eight steps, manipulation, therapy, informing, consultation, placation, partnership, power delegation and citizen control. But this ladder was not seen as

---

4 http://www.pentaho.com/.

adequate. Starting from previous works, Kingston [KIN 98] has proposed a six-step ladder (Figure 12.3) which appears more relevant to our purpose. Among the steps, one can successively find from bottom to top (the lower steps meaning no real public participation, and note that upper level correspond to different degrees of citizen empowerment):

– *public right to know*: in this first level, the public only has the possibility to be aware that some planning issue could be of interest;

– *informing the public*: here the concerned local authority implements some action plan in order to inform the public; but the public have no way of reacting;

– *public right to object*: here the city-dwellers may say yes or no to a project, but have no way of reacting nor can they amend it;

– *public participation in defining interests, actors and determining agenda*: this is the very first level of participation;

– *public participation in assessing consequences and recommending solutions*: now the public is truly involved in analyzing the impacts of possible decisions and can recommend solutions which can be accepted and implemented;

– *public participation in final decision*: this is real participation in the final decision; the decision is not only made by elected officers (city-councilors for instance), but each citizen can vote whether or not they accept the plan.

13.3.2. *Smart people and smart governance*

Three categories can be identified: citizens, local authority officers and politicians. According to Goodchild [GOO 07], citizens must be considered as sensors, but while others claim that public participation is the future, several are advocating for people empowerment. Empowerment of citizens supposes (*i*) that they have access not only to information, but overall to a system able to sketch the future, (*ii*) that they can express their opinion, and (*iii*) that their opinion is taken into account in the decision-making process.

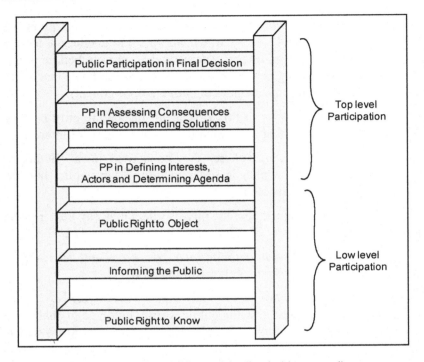

Figure 13.3. *The public participation ladder according to [KIN 98] with modifications*

The consequences are that local authority officers of Smart Cities must AUTOMATICALLY deliver information when requested by citizens; in other words, local information must no longer be classified. The second is that knowledge must be free to use. In other words, no companies must sell geographic knowledge.

We think that the key-issue of citizen empowerment is to let them use tools to develop their argumentation, to examine the consequences of their ideas and to simulate the future. In other words, the importance is not to make well-informed decisions only based on causes, but overall also on consequences. And consequences can be estimated or guessed by using geographic inference engines.

Figure 13.4 illustrates an example facing some vacant land (see also Rule 10.14). Some actors will advocate for a new recreation park whereas others for a commercial mall, etc. Sometimes the conflicts can be very strong and a consensus must be found in any way or some arbitration must close down the debate.

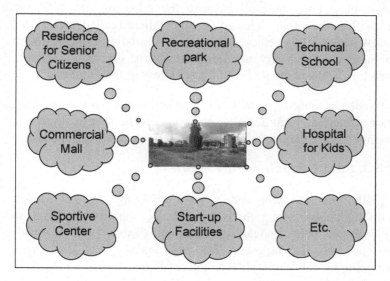

Figure 13.4. *Facing a vacant land, various stakeholders can have different visions*

The big problem regards general interest. Who is in charge of defining general interest? As NIMBY[5] interests can be easily assigned, it is more difficult for general interest: different people can define it differently. For some politicians, this notion is so vague that their interest is to be re-elected; as a consequence, they propose what they think useful to be re-elected and assume this is of general interest. Another aspect of this notion regards territorial coverage: local general interest can contradict the general interest at a higher level; for instance conflicts due to international transportation problems.

---

5 NIMBY = "Not in my backyard".

A nice example is taken from the TAV conflict in the Alps: the idea is to design a high-speed train between Lyon, France, and Turin, Italy. Both the Chambers of Commerce from Lyon and Turin together with local authorities are strongly in favor of this train which implies several interesting economic aspects not only for the region, but also at European levels overall because the Alps are a sort of natural barrier. On the other side, environmentalists (called NO-TAV) are against because this project because it will disfigure the landscape in the Suse Valley. Many manifestations were organized against this project. A solution would be a 50 km gallery in the mountain.

Finally, a smart city or territory must have smart politicians and smart officers as local decision-makers. It is the role of smart citizens not only to give their opinion but overall to elect smart politicians.

13.3.3. Smart people involvement

[MIE 08] has proposed a table showing the key-aspects of involving people along three dimensions, namely cognitive, socio-political and technological-organizational (Table 13.1) in which the last one was decomposed. We can observe that this vision is interesting by combining different disciplines.

Dimension	Components
Cognitive	Renovation of the paradigms for thinking out territorial development. Renovation of substantive theories on territorial development. Production of new empirical knowledge about contemporary territorial transformations.
Socio-political	New forms of vertical and horizontal coordination. New project territories. Territorial forecast and new collective projects.
Organizational	New agencies and institutional networks for territorial management.
Technological	New tool kit for analysis, monitoring and territorial communication.

Table 13.1. *Dimensions and new components required for territorial governance. Modified from [MIE 08]*

Therefore, in the perspective of the knowledge society, [MIE 08] asks for new tool kits, and knowledge engineering seems to be the key-issue for the background of such tools. And perhaps the most significant expression to summarize the role of knowledge in the world today is that attributed to the philosopher Michel Serres: "knowledge is now the infrastructure" [LEV 94].

13.4. Knowledge as infrastructure for smart governance

Assuming that knowledge is the infrastructure (Figure 1.8) the paramount questions are, (*i*) how to produce and organize it, (*ii*) how to make it accessible to citizens and (*iii*) how to use it efficiently?

13.4.1. *Geographic knowledge organization*

In Figure 3.4, a general structure of a geographic knowledge system was given. To this structure, several other components must be added:

– *documents and reports*; in which often not only information but also knowledge is hidden; the consequence is to transcend Geographic Information Retrieval for Geographic Knowledge Retrieval by creating new software products for this task;

– *external knowledge*; in Chapter 3, this kind of knowledge was rapidly studied, decomposed into neighboring and outside knowledge; as I have already advocated to include neighboring knowledge into GKS, it must be important to structure outside knowledge and especially outside good practices which could be of interest for smart governance;

– *previous projects*; especially the objective to store the knowledge gathered from them and the reasons for which they have succeeded or not;

– *current projects under development*; the different steps, the different alternatives, the criteria, the opinions of the stakeholders, etc must be stored.

13.4.2. *Geographic knowledge accessibility*

Not only experts, local authority officers, urban planners and politicians can be granted access to knowledge, but overall citizens, activists and any local associations must not be denied the access to knowledge. For home security, some chunks of information, knowledge or reports must be classified, it is not a reason to deprive access to anybody.

For technical reasons, since such a system may be complex, some facilitators can be appointed to help lay citizens.

13.4.3. *Categories of geographic knowledge*

In this book, we have encountered different types of knowledge:

a) *Consistency knowledge.* This is knowledge used to check the validity of any knowledge chunk when it is entered, updated or modified into the knowledge; for instance rules to check the validity of polygons, the validity of tessellations, etc. It also includes any rules for quality control.

b) *Mutation of geographic object's types and relations.* Those rules are general and are independent from location; they will ensure some robustness to geographic reasoning.

c) *Application knowledge.* Corresponding to issues, some bunches of knowledge can be regrouped. For instance, let us mention urban planning, transportation, energy, environmental planning, recreation parks, tourism, pollution, etc.

13.4.4. *Technology and sociological watching*

In order to increase the development of a smart territory, one interesting aspect is to try to benefit not only from new technologies, but also from others people's experiences. In other words, technology watching must be completed by a kind of sociological watching which integrates successful experiments made in other cities, in order to take innovations coming from both companies and other territories into account. Let us rapidly study each of them.

Figure 13.5. *Technology watch from http://www.biochem-project.eu/download/innova/newsletter/ktn/BIOCHEM_ktn_2012-03.pdf*

13.4.4.1. *Technology watch*

Technology watch's task is to observe, track, filter out and assess potential technologies from a very wide field extending beyond the normal confines of the sector (Figure 13.5). Often the expression "Technology Watching" is linked to "Competitive Intelligence".

For instance, the "humble" lamppost can be transformed into an intelligent sensor for light, air pollution, noise, etc.

13.4.4.2. *Urban sociological watching*

Another complementary aspect is to be aware of novel innovations in other cities and territories not only from a technological aspect but also from a sociological aspect. Facing a new transport-on-demand system established in city *A*, it could be interesting to analyze the outcomes in this city; and based on this analysis, to examine whether this solution can be imported to another city *B*. But overall, a local decision-maker must imagine how this kind of system can be integrated.

The spread of the biking system to several cities can be seen as a result of technological and sociological watching.

13.4.5. *Back on external knowledge*

As explained in section 3.4, geographic external knowledge can be useful to assert reasoning for territorial intelligence and smart cities. It can be decomposed into neighboring knowledge and outside knowledge.

13.4.5.1. *Neighboring knowledge*

Neighboring knowledge represents knowledge located at the vicinity of the jurisdiction, for instance within an out-buffer. In [CHE 15b], a rule of thumb proposes that the width of the out-buffer must be stated as $D = \sqrt{S} \ / \ 20$ in which S denotes the area of the territory.

It must include main geographic objects, relationships between those objects and the objects located inside the jurisdiction and especially cross-border rules.

13.4.5.2. *Outside knowledge*

Especially from technology and urban sociological watching, interesting experiments made in other territories or cities can be modeled and stored as external good practices. The primary step will be to analyze the semantics of outside knowledge and to propose a way or a language that can be machine-processable, for instance by a variant of case-based reasoning.

13.4.6. *Geographic knowledge on the move*

To run, update and manage such systems, some specialized administrators must be appointed. Their roles will be among others:

– to ensure the good functioning of the system;

– to check consistency regularly;

– to encapsulate what-if models to be integrated;

– to correctly manage sensors, maintain them, etc.;

– to watch modifications in neighboring knowledge;

– to model and integrate outside good practices;

– to work with facilitators for better assistance;

– to ease interoperability with other local authorities GKS;

– and to prepare new projects as explained in section 3.2.

13.5. Conclusion, from knowledge to wisdom

In Chapter 1, the pyramid data-information-knowledge-wisdom was presented. Now that the background of a knowledge infrastructure devoted to territorial intelligence and smart cities is sketched, the premises of wisdom are laid.

Beware of the meaning of "wisdom"! This is not wisdom in a philosophical point of view, but rather some coherence when applying knowledge. Consider Machiavelli's advice to govern, wisdom based on Machiavelli's advices can be perhaps interesting for a dictator, but is far from wisdom inspiring public participation!

What could wisdom be at this level? Many directions of work must be followed. Let us examine some of them:

a) First in accordance with applications, bunches of knowledge must be detected, for instance regarding transportation, energy management, pollution fighting, engineering network organization for electricity, water supply, for waste management, socio-cultural and sportive activities without forgetting risk and disaster management. The first step will be to write those rules in a human language, then when possible to translate them into a machine-processable language. The task to identify those rules is a heavy chore.

b) All mathematical models must be encapsulated in order to be used in such system. Maybe, some of them must be adapted in order to match data structures.

c) The characteristics and specifications of a geographic inference engine must be clarified and established.

d) More importantly, stakeholder's logic must be studied and specific rules must be written. Even if several monographs explain those logics, the rules must be identified.

e) Systems of arbitration or for finding consensus must be elaborated and integrated.

Once those preliminary works are fulfilled, the next problem is how local authorities will use such systems, not to be reelected, but for the happiness of people they are in charge of.

Is a new era coming from cities and territories?

Bibliography

[AGR 93] AGRAWAL R., IMIELINSKI T., SWAMI A., "Mining association rules between sets of items in large databases", *Proceedings of the ACM SIGMOD Conference on Management of Data*, pp. 207–216, 1993.

[AIR 16] AIRAKSINEN M., "Smart Cities, can the performance be measured?", *VTT Impulse*, 1/2016, pp. 26–33, VTT, Finland, 2016.

[AHL 94] AHLBERG, C., SHNEIDERMAN B., "Visual information seeking: tight coupling of dynamic query filters with starfield displays", *Proc. of ACM CHI94 Conference*, pp. 313–317, April 1994.

[ALB 15] ALBINO V., BERARDI U., DANGELICO R.M., "Smart cities: definitions, dimensions, performance and initiatives", *Journal of Urban Technology*, vol. 22, no. 1, pp. 3–21, available at: http://dx.doi.org/ 10.1080/10630732.2014.942092, 2015.

[ALE 64] ALEXANDER, C., *Notes on the Synthesis of Form*, Harvard University Press, Cambridge, 1964.

[ALL 83] ALLEN J.F., "Maintaining knowledge about temporal intervals", *Communications of the ACM*, pp. 832–843, 26 November 1983.

[ALT 95] ALT H., GODAU M., "Computing the Fréchet distance between two polygonal curves", *Int. Journal of Computational Geometry Applications*, vol. 5, no. 75, pp. 75–91, 1995.

[ANT 91] ANTENUCCI J.C., BROWN K., CROSWELL P.L. *et al.*, *Geographic Information Systems: A Guide to the Technology*, Van Nostrand Reinhold, 1991.

[ANT 12] ANTHOPOULOS L., VAKALI A., "Urban planning and smart cities: interrelations and reciprocities", in ÁLVAREZ F., CLEARY F., DARAS P. (eds), *Future Internet Assembly, from Promises to Reality*, Springer-Verlag, 2012.

[APP 03] APPICE A., CECI M., LANZA A. *et al.*, "Discovery of spatial association rules in geo-referenced census data: A relational mining approach", *Intelligent Data Analysis*, vol. 7, IOS Press, pp. 541–566, 2003.

[ARN 69] ARNSTEIN S.R., "A ladder for citizen participation", *Journal of the American Institute of Planners*, vol. 35, no. 7, pp. 216–244, 1969.

[BAD 09] BADARD T., DUBÉ E., "Enabling Geospatial Business Intelligence. The Open Source Business Resource", available at: http://geosoa.scg.ulaval.ca/~badard/article-tbadard-osbr_2009-long_version-enabling_geospatial_bi.pdf, September, 2009.

[BAT 91] BATTY M., YEH T., "The promise of expert systems for urban planning", *Computers, Environment and Urban Systems*, vol. 15, no. 3, pp. 101–108, 1991.

[BAX 75] BAXTER R.S., ECHENIQUE M.H., OWERS J. (eds), *Urban Development Models*, Construction Press, Lancaster, 1975.

[BAX 76] BAXTER R.S., *Computer and Statistical Techniques for Planners*, Methuen Press, 1976.

[BER 73] BERTIN J., *Sémiologie graphique*, 2nd ed., Mouton/Gauthier-Villars, 1973.

[BER 07] BERTACCHINI Y., RODRÍGUEZ-SALVADOR M., SOUARI W., "From territorial intelligence to competitive & sustainable system case studies in Mexico & in Gafsa University", *Second International Annual Conference of Territorial Intelligence*, Spain, pp. 37–54, October 2007.

[BER 12] BERTACCHINI Y., "Between information and communication process, the territorial intelligence, as a network concept & a framework to shape local development", *International Journal of Humanities and Social Science*, vol. 2, no. 18, October 2012.

[BER 16] BERTACCHINI Y., BOUCHET Y., "Territorial intelligence & artificial intelligence: on discussion", *Asian Journal of Computer and Information Systems*, vol. 04, no. 05, pp. 155–168, October 2016.

[BIA 16] BIASOTTI S., CERRI A., PATANÉ G. *et al.*, "Feature extraction and classification", in PATANÉ G., SPAGNUOLO M. (eds), *Heterogeneous Spatial Data: Fusion, Modeling, and Analysis for GIS Applications*, Morgan & Claypool Publishers, 2016.

[BLA 83] BLAKEMORE M., "Generalization and error in spatial databases", *Cartographica*, vol. 21, pp. 131–135, 1983.

[BOL 10] BOLEY, H., PASCHKE, A., SHAfiQ, O., "RuleML 1.0: the overarching specification of web rules", *Proceedings of Semantic Web Rules: International Symposium, RuleML 2010*, Washington, DC, October 21–23, 2010.

[BON 99] BONHOMME C., TREPIED C., AUFAURE M.A. *et al.*, "A visual language for querying spatio-temporal databases", *Proceedings of the 7th International Symposium on GIS, ACMGIS'99*, Kansas City, pp. 34–39, November 5–6, 1999.

[BOR 97] BORST W.N., AKKERMANS J.M., TOP J.L., "Engineering ontologies", *International Journal of Human-Computer Studies*, vol. 46, pp. 365–406, 1997.

[BOR 02] BORGES K.A.V., DAVIS JR., C.A. *et al.*, "Integrity constraints in spatial databases", in DOORN J.H., RIVERO L.C. (eds), *Database Integrity: Challenges and Solutions*, Idea Group Publishing, Hershey, 2002.

[BOU 17] BOUATTOU Z., LAURINI R., HAFIDA BELBACHIRA H., "Animated chorem-based summaries of geographic data streams from sensors in real time", *Journal of Visual Languages and Computing*, 2017.

[BRA 12] BRAVO L., RODRÍGUEZ A., "Formalization and reasoning about spatial semantic integrity constraints", *Data and Knowledge Engineering*, vol. 72, pp. 63–82, 2012.

[BRU 80] BRUNET R., "La composition des modèles dans l'analyse spatiale", *L'Espace géographique*, no. 4, 1980.

[BRU 93] BRUNET R., "Les fondements scientifiques de la chorématique", in "La démarche chorématique", Centre d'Études Géographiques de l'Université de Picardie Jules Verne, 1993.

[BUT 91] BUTTENFIELD B., MCMASTER R., *Map Generalization: Making Rules for Knowledge Representation*, Longman, London, 1991.

[CHE 15a] CHERNI I., FAIZ S., LAURINI R., "ChoreMAP: a new tool for extraction and visualisation for visual summaries", *7th International Conference on Information, Process, and Knowledge Management* (eKNOW'2015), Février, Lisbon, Portugal, 2015.

[CHE 15b] CHERNI I., Découverte de chorèmes par fouille de données spatiales, PhD Thesis, INSA of Lyon, September 11, 2015.

[CLE 93] CLEMENTINI E., DI FELICE P., VAN OOSTEROM P., "A small set of formal topological relationships suitable for end-user interaction", in ABEL D., OOI B.C. (eds), *Advances in Spatial Databases*, Springer, 1993.

[COH 96] COHN A., GOTTS N., "The 'Egg-Yolk' Representation of Regions with Indeterminate Boundaries", in BURROUGH P., FRANK A. (eds), *Geographic Objects with Indeterminate Boundaries*, Taylor & Francis, 1996.

[COM 14] COMIC L., DE FLORIANI L., MAGILLO P. *et al.*, "Morphological Modeling of Terrains and Volume Data", *Springer Briefs in Computer Science*, vol. XI, no. 116, 2014.

[CRO 99] CROWTHER P., The Nature and Acquisition of Expert Knowledge to be used in Spatial Expert Systems for Classifying Remotely Sensed Images, PhD Thesis, University of Tasmania, Australia, 1999.

[CUI 03] CULLOT N., PARENT C., SPACCAPIETRA S. *et al.*, "Des ontologies pour données géographiques", *Revue Internationale de Géomatique*, vol. 13, no. 3, pp. 285–306, 2003.

[DAT 87] DATE C.J., "Twelve Rules for a Distributed Database", *InfoDB*, vol. 2, nos. 2 and 3, 1987.

[DAV 98] DAVENPORT T.H., PRUSAK L., *Working Knowledge: How Organisations Manage What They Know*, Harvard Business School Press, Boston, 1998.

[DEC 11] DE CHIARA D., DEL FATTO V., LAURINI R. *et al.*, "A chorem-based approach for visually analyzing spatial data", *Journal of Visual Languages and Computing*, vol. 22, no. 3, June, 2011.

[DEL 07] DEL FATTO V., LAURINI R., LOPEZ K. *et al.*, "Potentialities of chorems as visual summaries of spatial databases contents", *9th International Conference on Visual Information Systems*, Shanghai, China, pp. 537–548, 28–29 June 2007.

[DEL 15] DEL FATTO V., DEUFEMIA V., PAOLINO L. *et al.*, "WiSPY: A Tool for Visual Specification and Verification of Spatial Integrity Constraints", *Journal of Visual Languages and Sentient Systems*, vol. 1, pp. 39–48, available at: http://ksiresearchorg.ipage.com/vlss/journal/DMS2015_proc_part2.pdf, August 2015.

[DIE 08] DIETZ J.L.G., "On the nature of business rules", *Advances in Enterprise Engineering*, vol. 10, pp. 1–15, August 2015.

[DJU 06] DJURIĆ D., GAŠEVIĆ D., DEVEDŽIĆ V., "The Tao of modeling spaces", *Journal of Object Technology*, vol. 5. no. 8, November-December 2006, pp. 125–147, available at: http://www.jot.fm/issues/issue_2006_11/ articlex, 2006.

[DOU 73] DOUGLAS D.H., PEUCKER T.K., "Algorithms for the reduction of the number of points required to represent a digitised line or its caricature", *The Canadian Cartographer*, vol. 10, no. 2, pp. 112–122, 1973.

[DOU 86] DOUGLAS D.H., "Experiments to locate ridges and channels to create a new type of digital elevation model", *Cartographica*, vol. 23, no. 4, pp. 29–61, 1986.

[EDW 13] EDWARDS P.N., JACKSON S.J., CHALMERS M.K. *et al.*, "Knowledge Infrastructures: Intellectual Frameworks and Research Challenges", Ann Arbor, Deep Blue, available at: http://hdl.handle.net/2027.42/97552, 2013.

[EGE 91] EGENHOFER M., FRANZOSA R.D., "Point-set topological spatial relations", *International Journal of GIS*, vol. 5, no. 2, pp. 161–174, 1991.

[EGE 94] EGENHOFER M., "Deriving the combination of binary topological relations", *Journal of Visual Languages and Computing* (JVLC), vol. 5, no. 2, pp. 133–149, June, 1994.

[EGE 05] EGENHOFER M., "Spherical topological relations", *Journal of Data Semantics*, vol. 3, pp. 25–49, 2005.

[EST 97] ESTER M., KRIEGEL H.-P., SANDER J., "Spatial data mining: a database approach", *Proceedings of 5th Symposium on Spatial Databases*, Berlin, Germany, 1997.

[EST 01] ESTER M., KRIEGEL H.-P., SANDER J., "Algorithms and applications for spatial data mining", in MILLEN H.J, HAN J. (eds), *Geographic Data Mining and Knowledge Discovery*, Taylor and Francis, 2001.

[FEL 99] FELLBAUM, C. (ed.), *WordNet: an Electronic Lexical Database*, MIT Press, Cambridge, 1999.

[FEN 97] FENSEL D., GROENBOOM R., "Specifying knowledge-based systems with reusable components", *Proceedings of the 9th International Conference on Software Engineering and Knowledge Engineering* (SEKE-97), Madrid, Spain, 1997.

[FER 16] FERNANDEZ-ANEZ V., "Stakeholders approach to smart cities: a survey on smart city definitions", *Proceedings of the First International Conference, Smart-CT* 2016, Málaga, Spain, pp. 42–51, June 15–17, 2016.

[FIS 13] FISTOLA R., LA ROCCA R.-A., "Smart city planning: a systemic approach", 6th *Knowledge Cities World Summit*, Istanbul, Turkey, pp. 520–529, September 9–12, 2013.

[FRI 97] NOY N.F., HAFNER C., "The state of the art in ontology design: a survey and comparative review", *AI Magazine*, vol. 18, no. 3, pp. 53–73, 1997.

[FUB 12] FU B., BRENNAN R., O'SULLIVAN D., "A configurable translation-based cross-lingual ontology mapping system to adjust mapping outcomes", *J. Web Semant*, vol. 15, pp. 15–36, 2012.

[GEN 06] GENG L., HAMILTON H.J., "Interestingness measures for data mining: a survey", *ACM Computing Surveys*, vol. 38, no. 3, pp. 1–32, 2006.

[GIR 08] GIRARDOT J.-J., "Evolution of the concept of territorial intelligence within the coordination action of the european network of territorial intelligence", *Ricerca e Sviluppo per le Politiche Sociali*, vol. 1, nos. 1–2, pp. 11–29, 2008.

[GIR 10] GIRARDOT J.-J., BRUNAU E., "Territorial intelligence and innovation for the socio-ecological transition", *9th International Conference of Territorial Intelligence*, ENTI, Strasbourg, France, 2010.

[GOL 02] GOLLEDGE R., "The nature of geographic knowledge", *Annals of the Association of American Geographers*, vol. 92, no. 1, pp. 1–14, 2002.

[GOM 02] GÓMEZ-PÉREZ A., CORCHO O., "Ontology languages for the semantic web", *IEEE Intelligent Systems*, vol. 17, no. 1, pp. 54–60, 2002.

[GOO 07] GOODCHILD M.F., "Citizens as sensors: the world of volunteered geography", *Geo Journal*, vol. 69, no. 4, pp. 211–221, 2007.

[GOR 01] GORDILLO S., Modélisation et manipulation de phénomènes continus spatio-temporels., PhD Thesis, Université Claude Bernard Lyon I, October, 2001.

[GRA 06] GRAHAM I., *Business Rules Management and Service Oriented Architecture: A Pattern Language*, John Wiley, London, 2001.

[GRU 93] GRUBER T.R., "A translation approach to portable ontologies", *Knowledge Acquisition*, vol. 5, no. 2, pp. 199–220, 1993.

[GRU 95] GRUBER T.R., "Toward principles for the design of ontologies used for knowledge sharing", *International Journal of Human-Computer Studies*, vol. 43, nos. 4–5, pp. 907–928, November 1995.

[GUA 98] GUARINO N., "Formal ontology and information systems", in GUARINO N. (ed.), *Formal Ontology in Information Systems*, IOS Press, Amsterdam, 1998.

[GUO 09] GUO D., MENNIS J., "Spatial data mining and geographic knowledge discovery – an introduction", *CEUS Computers, Environment and Urban Systems*, vol. 33, pp. 403–408, 2009.

[GUP 95] GUPTILL S.C., MORRISON J.L. (eds), *Elements of Spatial Data Quality*, Elsevier, Oxford, 1995.

[GUT 84] GUTTMAN A., "R-trees: a dynamic index structure for spatial searching", *Proceedings of the 1984 ACM SIGMOD International Conference on Management of Data – SIGMOD*, vol. 84, p. 47, 1984.

[HAL 08] HALATSCH J., KUNZE A., SCHMITT G., "Using shape grammars for master planning", *Design Computing and Cognition'08*, pp. 655–673, 2008.

[HAN 04] HAN Q., BERTOLOTTO M., WEAKLIAM J., "A conceptual model for supporting multiple representations and topology management", *Conceptual Modeling for Advanced Application Domains*, vol. 3289, pp. 5–16, 2004.

[HEC 13] HEĆIMOVIĆ Z., CICELI T., "Spatial intelligence and toponyms", in *Proceedings of the 26th International Cartographic Conference*, Dresden, Germany, 25–30 August 2013.

[HOL 02] HOLSAPPLE C.W., JOSHI K.D., "A collaborative approach to ontology design", *Communications of the ACM*, vol. 45, no. 2, pp. 477–490, 2002.

[ISH 05] ISHIBUCHI H., YAMAMOTO T., "Rule weight specification in fuzzy rule-based classification systems", *IEEE Transactions on Fuzzy Systems*, vol. 13, no. 4, pp. 428–435, August 2005.

[ITT 74] ITTEN J., *The Art of Color: The Subjective Experience and Objective Rationale of Color*, Wiley, 1974.

[JAI 11] JAIN K.P., "A review study on Urban planning and artificial intelligence", *International Journal of Soft Computing and Engineering* (IJSCE), vol. 1, no. 5, November 2011.

[JAK 11] JAKIR Ž., HEĆIMOVIĆ Ž., ŠTEFAN Z., "Names ontologies", in RUAS A. (ed.), *Advances in Cartography*, Springer Verlag, Heidelberg, 2011.

[JAN 08] JANOWICZ K., MAUÉ P., WILKES M. *et al.*, "Similarity as a quality indicator in ontology engineering", *Proceedings of the Fifth International Conference (FOIS 2008)*, Amsterdam, pp. 92–105, 2008.

[JEA 16] JEANSOULIN R., "Review of forty years of technological changes in geomatics toward the big data paradigm", *ISPRS International Journal of Geo-Information*, vol. 5, pp. 1–16, 2016.

[JOH 15] JOHANY F., BIMONTE S., "A framework for spatio-multidimensional analysis improved by chorems: application to agricultural data", *International Conference on Data Management Technologies and Applications*, pp. 59–80, 2015.

[JOS 10] JOOSTEN S., WEDEMEIJER L., MICHELS G., *Rule Based Design*, Open Universiteit, Heerlen, 2010.

[KAN 02] KANG M.A., FALOUTSOS C., LAURINI R. *et al.*, "Indexing values in continuous field databases", *Proceedings of the International Conference on Extending Database Technology* (EDBT 2002), Prague, March 2002.

[KAV 05] KAVOURAS M., KOKLA M., TOMAI E., "Comparing categories among geographic ontologies", *Computers & Geosciences*, vol. 31, no. 2, pp, 145–154, 2005.

[KEI 06a] KEITA A., ROUSSEY C., LAURINI R., "Un outil d'aide à la construction d'ontologies pré-consensuelles : le projet Towntology", *XXIVème Congrès INFORSID*, Hammamet, Tunisia, pp. 911–926, 2006.

[KEI 06b] KEIM D.A., MANSMANN F., SCHNEIDEWIND J. *et al.*, "Challenges in visual data analysis", *Tenth International Conference on Information Visualisation (IV'06)*, pp. 9–16, 2006.

[KEM 98] KEMP K.K., VCKOVSKY A., "Towards an ontology of fields", *Proceedings of the 3rd International Conference on GeoComputation*, September 1998.

[KEß 09] KEßLER C., JANOWICZ K., BISHR M., "An agenda for the next generation gazetteer: geographic information contribution and retrieval", *Proceedings of the 17th ACM SIGSPATIAL International Conference on Advances in Geographic Information Systems*, New York, NY, USA, pp. 91–100, 4–6 November, 2009.

[KIM 89] KIM T.J., WIGGINS LYNA L., WRIGHT J.R. (eds), *Expert Systems: Applications to Urban Planning*, Springer-Verlag, New York, 1989.

[KIN 98] KINGSTON R., "Web-based GIS for public participation decision making in the UK", *Proceedings of the Workshop on Empowerment, Marginalisation, and Public Participation GIS*, Santa Barbara, CA, available at: http://www.ccg.leeds.ac.uk/vdmisp/publications/sb_paper.html, October 14-17, 1998.

[KLI 05] KLIPPEL A., RICHTER K.-F., HANSEN S., "Wayfinding choreme maps", in *Proceedings of 8th International Conference*, (*VISUAL 2005*), pp. 94–108, June 2005.

[KOU 12] KOURTIT K., NIJKAMP P., "Smart cities in the innovation age", *Innovation: The European Journal of Social Science Research*, vol. 25, no. 2, pp. 93–95, 2012.

[LAC 76] LACOSTE Y., *La géographie, ça sert, d'abord, à faire la guerre*, Maspero, Paris, 1976.

[LAF 05] LAFON B., CODEMARD C., LAFON F., "Essai de chorème sur la thématiquede l'eau au Brésil", available at: http://webetab.ac-bordeaux.fr/Pedagogie/Histgeo/espaceeleve/bresil/eau/eau.htm, 2005.

[LAU 89] LAURINI R., MILLERET-RAFFORT F., *L'ingénierie des connaissances spatiales*, Hermès, Paris, 1989.

[LAU 93] LAURINI R., THOMPSON D., *Fundamentals of Spatial Information Systems*, Academic Press, 1993.

[LAU 98] LAURINI R., "Spatial multidabase topological continuity and indexing: a step towards seamless GIS data interoperability", *International Journal of Geographical Information Sciences*, vol. 12, no. 4, pp. 373–402, June 1998.

[LAU 00] LAURINI R., GORDILLO S., "Field orientation for continuous spatio-temporal phenomena", *International Workshop on Emerging Technologies for Geo-based Applications*, Ascona, Switzerland, May 22–26, 2000.

[LAU 01] LAURINI R., *Information Systems for Urban Planning: A Hypermedia Cooperative Approach*, Taylor and Francis, 2001.

[LAU 07] LAURINI R., "Pre-consensus ontologies and Urban databases", in TELLER J., LEE J.R., ROUSSEY C., (eds), *Ontologies for Urban Development*, Springer-Verlag, Heidelberg, 2007.

[LAU 12] LAURINI R., "Importance of spatial relationships for geographic ontologies", in CAMPAGNA M., DE MONTIS A., ISOLA F. *et al.* (eds), *Planning Support Tools: Policy Analysis, Implementation and Evaluation Proceedings of the Seventh International Conf. on Informatics and Urban and Regional Planning* (INPUT 2012), pp. 122–134, 2012.

[LAU 14] LAURINI R., "A conceptual framework for geographic knowledge engineering", *Journal of Visual Languages and Computing*, vol. 25, pp. 2–19, 2014.

[LAU 15a] LAURINI R., "Geographic ontologies, gazetteers and multilingualism", *Journal of Future Internet*, vol. 7, pp. 1–23, 2015.

[LAU 15b] LAURINI R., "Fundamentals of geographic engineering for territorial intelligence", in PEREZ GAMA A. (ed.), *Knowledge Engineering Principles, Methods and Applications*, Nova Science Publishing, New-York, 2015.

[LAU 15c] LAURINI R., "Primi passi per la modellazione delle regole geospaziali", ASITA Conference, Lecco, Italy, pp. 501–508, available at: http://atti.asita.it/ASITA2015/Pdf/323-662.pdf, 2015.

[LAU 16a] LAURINI R., KAZAR O., "Geographic Ontologies: Survey and Challenges", *Journal for Theoretical Cartography*, vol. 9, pp. 1–13, 2016.

[LAU 16b] LAURINI R., SERVIGNE S., FAVETTA F., "An introduction to geographic rule semantics", *Proceedings of the 22nd International Conference on Distributed Multimedia Systems (DMS 2016)*, Salerno, Italy, pp. 91–97, November 25-26, 2016.

[LBA 00] LBATH A., PINET F., "The development and customization of GIS-based applications and web-based GIS applications with the CASE tool AIGLE", *Proceedings of 8th ACM GIS*, Washington D.C., pp. 194–196, November 10–11, 2000.

[LEE 90] LEE S.-Y., HSU F.-J., "2D C-string: a new spatial knowledge representation for image database systems", *Pattern Recognition*, vol. 23, no. 10, pp. 1077–1087, 1990.

[LEE 92] LEE S.-Y., HSU F.-J., "Spatial reasoning and similarity retrieval of images using 2D C-string knowledge representation", *Pattern Recognition*, vol. 25, no. 3, pp. 305–318, 1992.

[LEJ 15] LEJDEL B., KAZAR O., LAURINI R., "Mathematical framework for topological relationships between ribbons and regions", *Journal of Visual Languages and Computing*, vol. 26, pp. 66–81, 2015.

[LEV 94] LEVY P., *L'intelligence collective. Pour une anthropologie du cyberspace*, La Découverte, Paris, 1994.

[LI 05] LI D., WANG S., "Concepts, principles and applications of spatial data mining and knowledge discovery", *International Symposium on Spatio-Temporal Modeling* (ISSTM 2005), Beijing, China, August 27–29, 2005.

[LI 15] LI D., WANG S., *Spatial Data Mining: Theory and Application*, 1st ed., Springer, 2015.

[LOP 06] LOPEZ K., Génération automatique de cartes chorématiques, Master Thesis, INSA of Lyon, June 2006.

[LOR 10] LORD P., "Components of an Ontology", http://ontogenesis.knowledge blog.org/514, 2010.

[MAC 97] MACEACHREN A.M., KRAAK M.J., "Exploratory cartographic visualization: advancing the agenda", *Computers & Geosciences*, vol. 23, no. 4, pp. 335–343, 1997.

[MAC 11] MACHADO I.M.R., ALENCAR R.O., CAMPOS JR. *et al.*, "An ontological gazetteer and its application for place name disambiguation in text", *Journal of the Brazilian Computer Society*, vol. 17, no. 4, pp. 267–279, 2011.

[MAT 13] MATHEW J., "City as a Customer", available at: http://www.frost.com/c/10046/ blog/blog-display.do?id=2377335, 2013.

[MEN 97] MENDELSON E., *Introduction to Mathematical Logic*, 4th ed., Chapman & Hall, London, 1997.

[MIE 08] MIEDES UGARTE B., "Territorial intelligence and the three components of territorial governance", *International Conference of Territorial Intelligence*, Besançon, France, p. 10, October 2008.

[MIL 09] MILLER H.J., HAN J., *Geographic Data Mining and Knowledge Discovery*, 2nd ed., CRC Press, 2009.

[MON 91] MONMONNIER M., *How to Lie with Map?*, The University Press of Chicago, Chicago, 1991.

[MOR 08] MORGAN T., *Business Rules and Information Systems: Aligning IT with Business Goals*, Addison-Wesley, 2008.

[MOU 16] MOURA T.H.V.M., DAVIS L.A. JR., FONSECA F.T., "Reference data enhancement for geographic information retrieval using linked data", *Transactions in GIS*, 2016.

[MUR 15] MURGANTE B., BORRUSO G., "Smart cities in a smart world", in RASSIA S., PARDALOS P. (eds), *Future City Architecture for Optimal Living*, Springer Verlag, Berlin, 2015.

[NEF 16] NEFZI H., FARAH M., FARAH I.R., "A similarity-based framework for the alignment of an ontology for remote sensing", *Computers & Geosciences*, vol. 96, no. C, pp. 202–207, November 2016.

[NON 95] NONAKA I., TAKEUCHI H., *The Knowledge Creating Company: how Japanese Companies Create the Dynamics of Innovation*, Oxford University Press, 1995.

[POP 34] POPPER K., *The Logic of Scientific Discovery*, 1934.

[PRE 85] PREPARATA F., SHAMOS M., *Computational Geometry: An Introduction*, Springer-Verlag, 1985.

[RAN 92] RANDELL D.A., CUI Z., COHN A.G., "A spatial logic based on regions and connection", *Proceedings of 3rd Int. Conf. on Knowledge Representation and Reasoning*, San Mateo, pp. 165–176, 1992.

[RAP 02] RAPER J., "The dimensions of GIScience", Keynote speech at GIScience, available at: http://www.soi.city.ac.uk/~raper/research/GIScience2002-OHs-pub.ppt, 2002.

[REI 09] REIMER A., DRANSCH D., "Information Aggregation: Automatized Construction of Chorematic Diagrams", available at: http://gfzpublic.gfz-potsdam.de/pubman/item/escidoc:239748:1/component/escidoc:239747/14094.pdf, 2009.

[ROS 11] ROSS R.G., "More on the if-then format for expressing business rules: questions and answers", *Business Rules Journal*, vol. 12, no. 4, available at: http://www.BRCommun 2002ity.com/a2011/b588.html, April 2011.

[SAL 13] SALLABERRY C., *Geographical Information Retrieval in Textual Corpora*, ISTE, London and John Wiley & Sons, New York, 2013.

[SAL 15] SALLEB-AOUISSI A., VRAIN C., CASSARD D., "Learning characteristic rules in geographic information systems", in BASSILIADES N., GOTTLOB G., SADRI F. *et al.* (eds), *Rule Technologies: Foundations, Tools, and Applications*, Springer, Berlin, 2015.

[SER 82] SERRA J., *Image Analysis and Mathematical Morphology*, Academic Press, 1982.

[SER 00] SERVIGNE S., UBEDA T., PURICELLI A. *et al.*, "A methodology for spatial consistency improvement of geographic databases", *GeoInformatica*, vol. 4, no. 1, pp. 7–34, 2000.

[SHE 94] SHEDROFF N., "Information Interaction Design: A Unified Field Theory of Design", available at: http://www.nomads.usp.br/documentos/ textos/design_interfaces_computacionais/info_interac_design_unified_nat han.pdf, 1994.

[SHE 01] SHEKHAR S., HUANG Y., "Discovering spatial co-location patterns: a summary of results", *Proceedings of the 7th Int'l Symposium on Spatial and Temporal Databases*, pp. 236–256, 2001.

[SHE 07] SHEKHAR S., ZHANG P., "Spatial Data Mining: Accomplishments and Research Needs", available at: http://www.spatial.cs.umn.edu/ paper_ps/giscience. pdf, 2006.

[SHE 16] SHEKHAR S., FEINER S.K., AREF W.G., *Spatial Computing*, *Communications of the ACM*, vol. 59, no. 1, pp. 72–81, 2016.

[SHN 97a] SHNEIDERMAN B., Information visualization: Dynamic queries, Starfield Displays and LifeLines, White Paper, available at: http://www. cs.umd.edu/hcil/members/bshneiderman/ivwp.html, 1997.

[SHN 97b] SHNEIDERMAN B., *Designing the User Interface*, 3rd ed., Addison-Wesley Publishing Company, 1997.

[SMA 07] SMART P.R., RUSSELL A., SHADBOLT N.R. *et al.*, "AKTiveSA: A Technical Demonstrator System For Enhanced Situation Awareness", *The Computer Journal*, vol. 50, no. 6, pp. 703–716, 2007.

[SMI 04] SMITH M.K., WELTY C., MCGUINNESS D.L. (eds), "OWL Web Ontology Language Guide", W3C Recommendation, available at: http:// www.w3.org/TR/owl-guide/, 2004.

[SOW 09] SOWA J.F., "Building, Sharing, and Merging Ontologies", available at: http://www.jfsowa.com/ontology/ontoshar.htm or http://www.jfsowa.com/talks/ontology.htm, accessed December 14 2015, 2009.

[SPI 95] SPIESS E., "The need for generalization in a GIS environment", in *GIS and Generalization: Methodology and Practise*, Taylor & Francis, London, UK, 1995.

[STI 78] STINY G., MITCHELL W.J., "The Palladian grammar", *Environment and Planning B*, vol. 5, pp. 5–18, 1978.

[STI 80] STINY G., "Introduction to shape and shape grammars", *Environment and Planning B: Planning and Design*, vol. 7, no. 3, pp. 343–351, 1980.

[STU 98] STUDER R., BENJAMINS V.R., FENSEL D., "Knowledge engineering: Principles and methods", *Data and Knowledge Engineering* (DKE), vol. 25, nos. 1–2, pp. 161–197, 1998.

[TEL 07a] TELLER J., "Ontologies for an improved communication in Urban development projects", in TELLER J. (ed.), *Ontologies for Urban Development*, Springer, 2007.

[TEL 07b] TELLER J., KEITA K., ROUSSEY C. *et al.*, "Urban ontologies for an improved communication in urban civil engineering projects: Presentation of the COST Urban Civil Engineering Action C21 TOWNTOLOGY", *European Journal of Geography Cybergeo*, vol. 386, 2007.

[THO 05] THOMAS J.J., COOK K.A., "Illuminating the Path R&D Agenda for Visual Analytics", National Visualization and Analytics Center (NVAC), US Dept of Homeland Security, available at: http://nvac.pnl.gov/agenda.stm, 2005.

[TOB 70] TOBLER W., "A computer movie simulating Urban growth in the detroit region", *Economic Geography*, vol. 46, no. 2, pp. 234–240, 1970.

[TUR 98] TURBAN E., ARONSON J.E., *Decision Support Systems and Intelligent Systems*, 5th ed., Prentice-Hall, 1998.

[VAN 97] VAN HEIJST G., SCHREIBER A., WIELINGA B., "Using explicit ontologies in KBS development", *Int. J. of Human-Computer Studies*, vol. 46, no. 2/3, pp. 183–292, 1997.

[VAR 16] VARADHARAJULU P., WEST G., MCMEEKIN D. *et al.*, "Automating government spatial transactions", in ROCHA J.G., GRUEAU C. (eds), *Proceedings of the 2nd International Conference on Geographical Information Systems Theory, Applications and Management* (GISTAM), Rome, Italy, pp. 157–167, 2016.

[VCK 95] VCKOVSKI A., "Representation of Continuous Fields", *Proceedings of 12th AutoCarto*, vol. 4, pp. 127–136, 1995.

[WEI 95] WEIBEL R., KELLER S., REICHENBACHER T., "Overcoming the knowledge acquisition bottleneck in map generalisation: the role of interactive systems and computational intelligence", in FRANK A.U., KUHN W. (eds), *Spatial Information Theory – A Theoretical Basis for GIS*, Springer, Berlin, 1995.

[YOO 14] YOO J.S., "Spatial data mining in the era of big data", available at: http://www.etcs.ipfw.edu/~lin/IEEE-FortWayneSection/TechMeeting/2014-5-6-Jin_Yoo_IEEE_FW-Talk.pdf, 2014.

[ZAD 65] ZADEH L.A., "Fuzzy sets", *Information and Control*, vol. 8, no. 3, pp. 338–353, 1965.

[ZAL 14] ZALA R.L., MEHTA B.B., ZALA M.R., "A survey on spatial co-location patterns discovery from spatial datasets", *International Journal of Computer Trends and Technology* (IJCTT), vol. 7, no. 3, available at: http://www.ijcttjournal.org, pp. 137–142, January 2014.

[ZHO 03] ZHOU S., JONES C.B., "A multi-representation spatial data model", in *Advances in Spatial and Temporal Databases*, Lecture Notes in Computer Science, vol. 2750, Springer , pp. 394–411, 2003.

Index

Printed in the United States
By Bookmasters